Scientific Epistemology

Scientific Epistemology

An Introduction

HILARY KORNBLITH

OXFORD
UNIVERSITY PRESS

OXFORD
UNIVERSITY PRESS

Oxford University Press is a department of the University of Oxford. It furthers
the University's objective of excellence in research, scholarship, and education
by publishing worldwide. Oxford is a registered trade mark of Oxford University
Press in the UK and certain other countries.

Published in the United States of America by Oxford University Press
198 Madison Avenue, New York, NY 10016, United States of America.

Library of Congress Cataloging-in-Publication Data
Names: Kornblith, Hilary, author.
Title: Scientific epistemology : an introduction / Hilary Kornblith.
Description: New York, NY, United States of America :
Oxford University Press, [2021] |
Includes bibliographical references and index.
Identifiers: LCCN 2021016356 (print) | LCCN2021016357 (ebook) |
ISBN 9780197609552 (hb) | ISBN 9780197609569 (paperback) |
ISBN9780197609583 (epub) | ISBN 9780197609590
Subjects: LCSH: Philosophy and science. | Science—Philosophy. | Knowledge, Theory of.
Classification: LCC B67 .K67 2021 (print) | LCC B67(ebook) | DDC 121—dc23
LC record available at https://lccn.loc.gov/2021016356
LC ebook record available at https://lccn.loc.gov/2021016357

DOI: 10.1093/oso/9780197609552.001.0001

1 3 5 7 9 8 6 4 2

Paperback printed by Marquis, Canada
Hardback printed by Bridgeport National Bindery, Inc., United States of America

To my students

Contents

Preface

When I was first approached about the possibility of writing an introductory book on the theory of knowledge, I immediately thought of a certain standard format for such books. One attempts to provide an overview of the field, surveying the different issues which are addressed by current theories, providing a guide to the strengths and weaknesses of the available approaches. Such an overview can be presented in a way which attempts to be as neutral as possible between the various competing positions, or it can take a stand, straightforwardly attempting to defend the merits of one particular view rather than others. Books of this sort, whether avowedly neutral or frankly partisan, are exceptionally useful. They introduce readers to the lay of the land, and readers come away with an understanding of the current state of the art.

The more I thought about this, the more clear it was to me that I didn't want to write such a book. There is a simple reason for this: there are already a great many such books introducing the theory of knowledge both to philosophy students and to interested readers outside a university setting, and, in my view, there are very many such books which do this extremely well.[1] There's little point then in writing another such book when I would merely be reinventing the wheel.

It occurred to me, however, that there is room for a very different sort of introduction to the field. Instead of trying to present the state of the art, covering the ins and outs of rival approaches, I could simply present my own view of how the theory of knowledge ought to be approached, making clear how my own approach offers an illuminating view of some issues worth thinking about. This would not just be an opinionated overview of the field, for I would not even attempt to survey all of its rivals. I would simply present one way of thinking about some issues in the theory of knowledge, making the case for that view as best I can.

I was buoyed in thinking that this might be worthwhile when I noted that the standard introductions to the field—introductions I think very highly of—often give little space to the kind of view I favor, and when they do leave space for my preferred approach, they often present it—at least as I see it—unsympathetically. In addition, when I think back to books I read as an undergraduate which offered introductions to various fields that I found intellectually exciting, they were often books of this sort: not neutral surveys of available views, or even opinionated surveys of available views, but frankly partisan approaches laying out one way in a which a field might be pursued which the author championed with full-throated enthusiasm.

It is important when writing such a book to be straightforward about what one is doing. Readers need to know that they are getting one way of looking at things when other no less responsible authors would present a very different view. So let the reader be warned: I am not even trying to survey the field here. My view is that the theory of knowledge should be informed by a scientific understanding of cognition, and that approach, which many philosophers would reject, is pursued throughout this book. I am very far from the only epistemologist who favors such an approach, but this book does not attempt to present only views which are a matter of consensus.

I should add, as well, that a book like this is best read alongside other books which present different views, whether by surveying the field, or, instead, like this book, by presenting some single way of viewing matters. I said that as an undergraduate, I especially appreciated reading one-sided introductions to various fields, but that was not because I thought that ignorance of the range of possible views on some subject was the best way to understand it. One needs to have some understanding of the different views that are available if one is to properly understand any of them. One need not gain that understanding, however, from any single book. Don't let this be the only thing you read about the theory of knowledge. I do hope, however, that you will find this a stimulating introduction to one way of thinking about what a theory of knowledge might be.

One more thing before we begin. You will notice that I have provided endnotes rather than footnotes to the text, and you may wonder why this book is formatted in such an inconvenient manner, forcing you to flip back and forth from the text to the end of the volume. This is intentional. The endnotes provide additional information and references for those who wish to see various complexities which the body of the text leaves out, or still more readings which go beyond the suggestions for further reading at the end of each chapter. Many readers will have little desire to pursue such matters, but even for those who do, these things can be a distraction. Better, to my mind, to tuck them away at the end of the book rather than distract the great majority of readers who wish to follow the main lines of the argument undistracted.

Acknowledgments

This book would never have been written were it not for John Fischer. John suggested that I write an introductory epistemology book, and he encouraged me in the thought that a book of this sort—by no means the standard sort of introductory book—would be worthwhile. As if that were not enough, it is by virtue of his support that this book appears with Oxford University Press.

Peter Ohlin at Oxford has been a wonderfully supportive editor. I am deeply grateful for all his work in seeing this project through. Referees for Oxford have been generous with their time and their constructive comments; I have benefited from their input.

My sister, Regina Kornblith, read and commented on chapters of the book as I wrote them, often providing me with wonderfully useful suggestions on the very day she received the manuscript. The book is better for all her advice.

I have received helpful feedback over many years from countless audiences at conferences and department colloquia, as well as from many friends and colleagues in the broader community of epistemologists. Their help and encouragement not only have contributed tremendously to my work, but have made it a pleasure.

The ideas in this book have been tried out in the classroom year after year, undergoing numerous revisions and additions. The students who attended those classes and who contributed so actively to discussion have my profound thanks. In addition to classroom work, my close work with graduate students has been invaluable in contributing to my education and to making my professional life so rewarding. It is for these reasons that I dedicate this book to the many students I have worked with over the course of my career.

Scientific Epistemology

1

The Threat of Skepticism

1.1. What Is a Theory of Knowledge and Why Do We Need One?

This is a book about the theory of knowledge, also known as epistemology, but we need to begin by explaining what a theory of knowledge is and why we need one.

Other disciplines don't face the same problem. There's no problem at all in explaining what a theory of motion is and why we need one. Physical objects move through space, and it would be nice to know if there are some regularities to be found about how they move, something more informative than just the familiar principle that what goes up must come down. As it turns out, there are such regularities. A theory of motion tells us what the laws of motion are which govern moving objects. Why do we need such a theory? One reason might be simple intellectual curiosity. The world is a complicated place, and we might be interested in learning more about how it works. This is a perfectly good reason to want a theory of motion. But there are obvious practical reasons as well. If a stone is thrown into the air, one might have very practical reasons for wanting to know where it will land. If a large meteor is headed toward the earth, there are straightforward practical reasons for wanting to know whether it will collide with the earth or harmlessly pass it by. And so on. As a result, when physicists talk about a theory of motion, no one is terribly puzzled about what such a theory might be or why anyone needs such a theory.

The same seems to be true about scientific theories generally. No one is terribly puzzled about what a theory of disease might be or why we need one. Or a theory of electricity. Or a theory of inheritance. And so on.

But philosophical theories are different, and there seem to be no such obvious answers to the questions about what a theory of

Scientific Epistemology. Hilary Kornblith, Oxford University Press. © Oxford University Press 2021.
DOI: 10.1093/oso/9780197609552.003.0001

knowledge might be or why we might need one. One straightforward way to address these issues is to ask what has made philosophers think they needed a theory of knowledge, and what they have thought such a theory might look like. And from the very beginnings of Western philosophy, in ancient Greece, the motivation for developing a theory of knowledge has come from the threat of skepticism.

Skepticism is the view that we know nothing at all. On its face, the view seems to have little to recommend it. We seem to know a great many things, from the utterly trivial, such as who is buried in Grant's tomb, to the highly consequential, such as whether it is safe to travel by airplane. The reason for taking skepticism seriously is not, therefore, that it seems to be true; it doesn't seem to be true. From the earliest days of philosophical thought, however, there have been various arguments which seemed to show that knowledge is, in fact, impossible. These arguments present us with a puzzle: they start from premises which seem undeniable, and move, by way of a series of inferences which themselves seem rationally forced upon us, to a conclusion which seems obviously false, the conclusion that we know nothing at all. These arguments present us with a problem: if we really do have lots of knowledge, as common sense would have it, and skepticism is thus false, then there is something wrong with these arguments, and we ought to be able to say what that is. Alternatively, if we cannot find any fault in these arguments, if they really do start with premises which are undeniable and move by way of a series of inferences which are themselves rationally forced upon us, then perhaps we need to take skepticism more seriously than common sense would suggest. And this has turned out to be a difficult problem because it has proven exceptionally difficult to say just where the error lies in various arguments for skepticism.

Skeptical arguments thus, historically, have provided a motivation for theorizing about knowledge.[1] We want a theory of knowledge because we want to be able to explain, in the face of skeptical arguments, how knowledge is actually possible. A theory of knowledge would provide us with an account of what knowledge is, and this, in turn, would allow us to respond to the threat of skepticism.

This isn't, of course, a practical problem. There aren't skeptics standing around, lurking on every corner, demanding that we explain how knowledge is possible and threatening us with some concrete harm if we cannot meet the demand. But it is an intellectual problem, and no less legitimate for all that. We all believe that we have a good deal of knowledge, and skeptical arguments seem to show that we are wrong about that. We ought to be able to respond to the challenge which these arguments present. We ought to be able to explain how knowledge is possible in the face of arguments which seem to show that it isn't. And the first thing we will need, it seems, if we are to successfully respond to any such argument is an account of what knowledge is. That would be a theory of knowledge.

If we are to make any progress here, we will thus need to look at some of these skeptical arguments and see why they present such difficult problems. So let us turn to one of the most powerful and troublesome skeptical arguments, the argument from illusion.

1.2. The Argument from Illusion

Right now I am sitting in a chair in my office. There are some books in front of me, a desk, a telephone, a computer, a coffee cup. All of these things are in plain view; I can see them clearly. I have no doubt whatsoever that I am in my office, rather than, say, at home. I have no doubt that there are books in front of me, and the various other objects which form the tools of my trade: my desk, my telephone, my computer, my coffee cup. In a word, I know that I am in my office; that there are books, and a desk, and a telephone, and a computer, and a coffee cup all within arm's reach. Nothing could be plainer than this. I can just see these things. To put the point somewhat pedantically, I am currently having certain sensory experiences on the basis of which I believe a host of things about where I am and what is in front of me. If I didn't have these sensory experiences, if I didn't, in particular, have the visual experiences I'm currently having, I wouldn't have these beliefs about my environment. It is my visual experience which provides my basis for believing what I do about my immediate surroundings. More than this, my visual experience provides me with exceptionally good reason to believe as I do. This is not a matter of wishful thinking or hasty generalization or some failure to consider relevant evidence. My visual experience provides all the evidence I need for believing as I do. It is on the basis of my visual experience that I may be credited with knowledge of the various features of my office. This isn't anything momentous, but it is knowledge nonetheless.

I could, if I wanted to, get additional evidence for these beliefs. I could, for example, reach out and touch these various objects to further confirm what my visual evidence tells me. In fact, I do, quite regularly, reach out and pick up my coffee cup. But even before I touch my coffee cup, and even though I haven't touched my telephone or various books, the evidence which my visual experience provides is evidence enough. I have excellent reason to believe as I do about the objects in my office even though I could have additional reasons for such beliefs. And it is because I have such good evidence and, of course, because my beliefs are, in fact, true, that I count as having knowledge about the objects in my office.

So what is the problem? Why does the skeptic think that I don't know these things? Indeed, why is it that the skeptic claims not only that, in fact, I don't know these things, but that knowledge of these things, or any other things at all, is impossible? The skeptical argument—in this case, the argument from illusion—is distressingly simple.

Consider my knowledge of my beloved coffee cup. I've had this coffee cup for several decades. It's black, with penguins on it, some of whom are wearing bow ties. I'm looking right at it now and that's how I know that my coffee cup is in front of me. The skeptic will point out, however, that I could have had the very visual experience I'm having right now—the experience on the basis of which I've come to believe that my coffee cup is in front of me—even if my coffee cup were not in front of me. People sometimes have hallucinations which are extraordinarily lifelike. If someone slipped some hallucinogenic drug into my breakfast cereal, I might have a visual hallucination of a coffee cup, a coffee cup just like mine, with well-dressed penguins on it, even when there is no coffee cup in front of me. And we can imagine a situation in which I have such an hallucination, without realizing that I'm hallucinating. Everything might seem perfectly normal to me. It might seem to me that I'm looking at my coffee cup in my office, just the way I do each morning, even though, in fact, I'm not looking at my coffee cup, but hallucinating.

Alternatively, the skeptic might point out, as the seventeenth-century French philosopher René Descartes did, that all of us have dreams. There is a very common anxiety dream which students often have. I myself had such dreams on more than one occasion when I was a student. I would dream that I was walking across campus, without a care in the world, when it suddenly occurred to me that I was due at a final exam. In one version of the dream, I also realized that I had no idea where the final was being held. In another, I realized that I had not attended even one meeting of the course all semester, nor had I so much as looked at any of the assigned readings. In another version—not that I ever had this version of the dream—one realizes as one is walking across campus that one is not wearing any clothes at all. In each case, panic sets in. And then one wakes up in a sweat.

Why do we experience such anxiety in these dreams? We do so because, when we're dreaming, we form a variety of beliefs: that we're walking across campus; that we should be taking a final exam; that we have no idea where the exam is being held; and so on. And we have these beliefs as a result of various experiences we have while dreaming. I might believe, for example, that I'm on campus walking past the library because I have certain visual experiences while dreaming, the very experiences I would have if I were on campus walking past the library. Our dreams can be a source of various emotions which we vividly continue to experience when we first awaken precisely because those experiences are so lifelike, that is, because, in dreams, we have experiences, like my visual experience as of walking past the library, that are exactly like experiences we have when awake.

The recognition that our sensory experience in dreams can be exactly like the sensory experience we have when awake led Descartes to ask himself what reason he has to believe, at any particular time, that he is in fact awake. So let us go back to my coffee cup. I'm in my office, as I said, looking right at the cup. And I believe, of course, that I'm looking at my cup because I can see it; that is, I believe the cup is right in front of me on the basis of the visual experience I currently have. But we now recognize that I could have experiences exactly like this in a dream. I could dream that I'm sitting in my office looking at my coffee cup. And if I did have such a dream, my visual experience would be exactly like the experience I'm having right now. And this raises the question: What reason do I have, given this fact, to believe that I am in fact awake now, in my office, looking at my coffee cup rather than asleep, at home, in my bed, with my eyes closed, and no coffee cup anywhere near me?

We've already said that I believe that my coffee cup is in front of me on the basis of my visual experience. But I could have an experience exactly like the one I'm having now even if I weren't looking at the cup and even if there were no cup present. I could have this very experience if I were asleep and dreaming. So, it seems, I have no more reason, at least on the basis of my experience, to believe that I'm awake rather than dreaming, or to believe that there's a coffee cup here rather than that there isn't. Nothing in my experience could possibly settle this issue, because for any experience one might pick, and for any feature of my experience that might serve as a sign that I am genuinely awake, I could have an experience exactly like that while dreaming.

Thus, consider the famous pinch test. How do I know I'm not dreaming now? I pinch myself, and I don't awaken; everything remains as it was before, and it does so because I was wide awake the whole time. Now this is, of course, a ridiculous test. It doesn't help at all. I could dream that I pinched myself and dream that I passed the test. But what applies to the pinch test applies to any would-be test to distinguish dreaming from waking life: I could dream that I applied the test and dream that I passed it. And this seems to show, as Descartes suggests, that I can't know right now that I'm actually awake rather than asleep and dreaming. And if I can't know that I'm awake rather than asleep and dreaming, then I can't know that I'm currently looking at a coffee cup, and so I can't know that there's a coffee cup in front of me. And if I can't even know that there's a coffee cup in front of me when I'm looking right at it, then I can't know anything at all about the world around me. There's nothing special about my knowledge of the coffee cup. Indeed, we began with this example because it seems so unproblematic. If it turns out that I can't even know this, then there is nothing at all I can know. Skepticism is forced upon us.

Descartes raises another skeptical worry, one even more fanciful. He asks us to imagine that there is an evil demon who is all-powerful and who wishes to deceive me about absolutely everything. The demon gives me visual experiences as of looking at my coffee cup when, in fact, there is no coffee cup in front of me. He gives me visual experiences as of being in my office when, in fact, I am not in my office; in fact, I have no office. He gives me apparent memories as of growing up in New York when, in fact, I didn't grow up in New York; in fact, there is no New York. And so on. And Descartes then asks what reason we could possibly have for thinking that there is no such demon. Just as in the dream case, nothing in our experience could possibly rule this out. For any experience we might choose to distinguish veridical experience—experience which accurately reflects the world around us—from illusory experience, the demon, it seems, could give us experience like that in order to mislead us. And that surely shows, Descartes suggested, that we are not in a position to know that our current experience is veridical and that we are not being deceived by a demon at this very moment, and that, in turn, undermines the basis for everything we believe about the world around us. We are thus led, inevitably, to total skepticism.

A more contemporary version of Descartes's evil demon involves a mad neurophysiologist. The neurophysiologist has you kidnapped and brought to his laboratory. Your brain is removed from your skull and put in a vat of nutrients to keep it functioning, and it is then provided with electrical stimulation exactly like the stimulation that would be supplied, for example, to the optic nerve, and to other sources of sensory input, were you, say, on campus, walking past the library. A whole simulated life is then provided for you. (Fans of *The Matrix* will have no problem imagining this sort of thing.) Once again, we see that there is nothing we can point to in our experience that could possibly show that we are currently embodied and in touch with the world around us rather than envatted in the mad neurophysiologist's laboratory. And, once again, this undermines the basis for everything we believe about the world around us. And, once again, we are inevitably led to total skepticism.

These nightmare scenarios, as I will call them, all present situations in which even our most confidently held beliefs are false. I'm looking right at my coffee cup now. I know that there is a coffee cup in front of me as surely as I know anything. But in each of these nightmare scenarios—the hallucination case, the dreaming case, the evil demon, the dastardly neurophysiologist—I have the very same visual experience I'm having right now, the same experience on the basis of which I've come to believe that there is a coffee cup in front of me, and yet my belief is false. What this suggests is that my current visual experience does not provide an adequate basis for believing that there is a coffee cup in front of me. Does that mean that I should hold off on believing that there is a cup in front of me until I can get additional evidence? Not at all. In each of these nightmare scenarios, there is simply no possibility of getting the kind of evidence that would allow me to determine whether I'm actually looking at the coffee cup right now or, instead, hallucinating, or dreaming, or being deceived by a demon. For all my experience shows, I could be a disembodied brain floating in a vat of nutrients, subjected to a steady stream of electrical stimulation which presents a thoroughly inaccurate picture of the world outside my vat. So there's no further evidence I can get which would allow me to determine whether there really is a cup in front of me. And what that seems to show is not only that I don't currently know whether there is a coffee cup in front of me, but that I couldn't possibly know such a thing. And this is not, of course, something peculiar to my belief about the coffee cup. Rather, these nightmare scenarios seem to demonstrate that I have no knowledge at all, and that it is absolutely impossible for me to gain any knowledge at all. That is the skeptical conclusion.

One last point about these arguments. You might think that perhaps the skeptical conclusion isn't so bad. If these arguments show what they seem to show, that we have no knowledge at all, and that knowledge is impossible to achieve, I should stop claiming to know things, and I should stop believing that I know things. That would certainly require me to change what I say and what I think. But maybe this wouldn't be as big a change as it seems. I should stop talking about knowledge, but life could go on as before. In particular, I still have just as much reason to behave as I do right now. I just have to get used to giving up all claims to knowledge.

But the skeptical arguments, if they work at all, show far more than this. Consider the following case. I'm standing on the corner of University Avenue intending to cross the street. Just as I'm about to set foot in the street, I see a large eighteen-wheel truck barreling straight toward me. Before thinking about the skeptical arguments, I would have said that I know there is a large truck headed straight toward me, and I know that it would be dangerous, even foolhardy, to cross the street. But I'm wiser now, and I realize that I shouldn't claim to know such things. I lack all knowledge of such things; knowledge is impossible. But I still, one might think, have reason not to cross the street. Even if I can't know that there is a truck headed toward me, surely I have good reason to believe that the truck will run me down if I walk into the street. And that's all I need to stay on the sidewalk. Life without knowledge goes on because my actions can be governed by reasonable belief even if knowledge is impossible. And reasonable belief is all it takes to keep me out from under the wheels of the oncoming truck.

But this attempt to endorse the skeptical argument while limiting the scope of its consequences cannot possibly work. What the skeptical argument shows, if it succeeds at all, is not just that knowledge is impossible. It shows that we have no more reason to believe that there is a truck headed down the street than that we are hallucinating that there is such a truck. We have no more reason to believe that there is a truck barreling down the street than that we are merely dreaming that there is such a truck. We have no more reason to believe that there is a truck about to crush us than that we are deceived by an evil demon, or a mad neurophysiologist, into thinking that there is such a truck when, in fact, there really isn't. And what all of this would mean, of course, is that we have no more reason to stay on the sidewalk than to walk boldly out into the street.

The skeptical conclusion is thus no small matter. It cannot be accommodated by some little change in our lives in which we stop talking about knowledge or thinking that we know things, and yet go on just as we did before. If the skeptical conclusion is true, not only do we know nothing at all, not only is knowledge impossible, but we have no reason whatsoever to do any of the things we do rather than anything else. We have no more reason to stay out of the way of oncoming trucks than we do to walk right in front of them. And that, if it were true, would matter a great deal.

I'm not suggesting for a moment that skepticism is true. I don't believe it. I certainly hope you don't believe it. Even Descartes didn't believe it. He thought he had a solution to the skeptical problem, an adequate reply to the skeptical argument. But I hope it is now clear why there is a problem about skepticism. We have an argument before us which seems to show that knowledge is impossible. And it isn't remotely clear what is wrong with this argument. It would be nice to be able to say just where the argument goes wrong. What that would take is a theory of knowledge, an account of what knowledge is which would show us how knowledge is genuinely possible. And that is one reason why philosophers engage in epistemological theorizing.

1.3. Convincing the Skeptic

What would it take to have an adequate reply to the skeptical argument? One very natural thought here is this. We imagine being confronted by someone who is convinced by the skeptical argument. Let us call this person the Skeptic. The Skeptic lays out the various nightmare scenarios for us and concludes that he has no knowledge whatsoever, and neither does anyone else. Knowledge, he tells us, is impossible. And now we try to devise an argument to convince the Skeptic that he is wrong: not only is knowledge possible, but we all do know many, many things. More than this, we cannot convince the Skeptic to change his mind by way of threats, or financial incentives, or any other inducements. We need to offer the Skeptic reasons to believe that skepticism is false. We succeed in responding to skepticism when we have devised a good argument which convinces the Skeptic that knowledge is possible and we do know many things. This is what an adequate response to skepticism requires, or so it might seem.

In order to see how such an argument might go, it will be useful to begin by considering how we actually convince others with far less radical views than the Skeptic. After all, we regularly encounter people who disagree with us about matters large and small, and such disagreement often prompts a kind of rational engagement: we each try to convince the other that we are right and they are wrong. Sometimes we succeed, and they are convinced by us; sometimes they succeed, and we are convinced by them; and sometimes there is a standoff: neither party to the dispute is convinced to change their mind.

Let us take an example of such a dispute. Suppose I meet someone who is a climate change skeptic: this person does not believe that there is human-caused global warming, while I do. I think this is an important issue, and I attempt to change this person's mind. I don't offer threats or incentives to believe as I do; I offer rational arguments.

Perhaps I begin by summarizing the scientific consensus: the overwhelming majority of climate scientists, I point out, believes that there is global warming and that it is caused by human activity. These experts know far more about this matter than I do or, as I suggest, than my opponent does, and in the face of such expertise, we should believe what the experts believe.

Now it might be nice—indeed, surely it would be nice—to explain what the reasons are on the basis of which the experts have come to this opinion. Nevertheless, it does seem that the argument I offer is a perfectly good reason to believe that global warming is indeed a real phenomenon. Understanding why global warming is occurring requires more than this, but this is sufficient reason to believe that global warming is actually occurring.

Nevertheless, this might not convince my opponent, and for perfectly good reason. I am happy to defer to the scientific consensus because I believe that climate scientists are better informed about the matter than I am, or, for that matter, than my opponent is. But my opponent might disagree with me about this. My opponent may know perfectly well what the scientific consensus is, but believe that climate scientists are just mistaken. Indeed, my opponent may have an explanation for why so many scientists might have all gone wrong: they are part of a liberal elite, my opponent claims, and they are driven by ideology rather than by the facts.

Now I may believe—in fact, I do believe—that this view of climate scientists is completely mistaken. But if I'm trying to convince someone who believes in such a vast liberal conspiracy that global warming is genuine, it will do me no good to take as one of my premises something which my opponent doesn't believe. If I were entitled to do that, then I might as well have simply stated my conclusion, that global warming really is occurring, and let it go at that. But this would obviously give my opponent no reason at all to believe as I do.

We can draw a general moral from this: if my argument is to be dialectically effective against an opponent, that is, if my argument is to give my opponent good reason for changing their view, then I cannot simply assume something which my opponent rejects. And that means that I cannot take as a premise in my argument a claim, such as the claim that we should defer to the scientific consensus, which my opponent does not accept. What we need to do, if we are to rationally convince an opponent, is to start with things that person already believes, and show them that these beliefs which we share rationally force them to change their mind about the issue which divides us. Our argument must therefore begin by getting a foothold within their body of beliefs. Without that, we offer our opponent no reason at all to change their mind.

But it takes more than premises to make an argument. Suppose I point out that the *New York Times* has reported something about the global warming issue, and I immediately conclude from this that matters are just as the *Times* has reported. Perhaps this is just an obvious conclusion to draw, as I see it: the *Times* says it; that settles it. Now my opponent may completely agree with me about what the *Times* has reported; I am then entitled to take that as a premise in my argument. But I have directly inferred something from this—that the facts are as the *Times* reports—without the benefit of further premises. I have treated this as a legitimate inference. Here too, my argument will not be dialectically effective—it will not rationally convince my opponent—if my opponent does not already see this as a legitimate inference. If my opponent rejects this inference, that does not mean that I cannot convince them. I may produce additional premises, premises which my opponent accepts, and make use of transitions which my opponent sees as legitimate, to show that the *Times* should be deferred to. But absent such additional argument, if my opponent rejects my inference, I have not produced the kind of argument which is needed to produce a rational change in view.

The kind of argument I need, then, to rationally change someone's mind must have premises which my opponent already accepts, and move by a series of inferences which my opponent takes to be legitimate. It is only in this way that we can show someone, given what they are already committed to, that they need to change their mind about some issue. This is what a dialectically effective argument requires.

Let us now leave the global warming skeptic behind and return to the Skeptic, that is, the person who claims that knowledge is impossible. In order to rationally convince this person that knowledge is, in fact, possible, and that we do know many things, we must provide an argument which starts with premises which the Skeptic already accepts. But what premises are these? The Skeptic insists that he knows nothing at all. So he doesn't accept any premises. Any attempt to provide premises from which our argument might proceed will automatically be dialectically ineffective against the Skeptic because we cannot get a foothold within the Skeptic's body of beliefs. And there is a very simple reason for this: the Skeptic does not have a body of beliefs; the Skeptic doesn't believe anything at all.[2]

Not only does the Skeptic have no beliefs; the Skeptic also does not accept any inferences as legitimate. To accept that an inference is legitimate would be to acknowledge that a certain transition in thought is sufficient to transmit knowledge. But the Skeptic, being a skeptic, doesn't accept that there are any such legitimate transitions.

So, to summarize: in order to rationally convince the Skeptic that knowledge is, indeed, possible, and that we do have a great deal of knowledge, we need to construct an argument which leads to that conclusion starting from premises the Skeptic already accepts and proceed by way of a series of inferential transitions which the Skeptic acknowledges as legitimate. But the Skeptic has no beliefs, and so there are no premises which the Skeptic already accepts, and the Skeptic does not accept any inferential transitions as legitimate. So we need to provide an argument for the conclusion that knowledge is possible which has no premises and involves no inferences. And, of course, there are no such arguments. So it is clearly impossible to come up with a dialectically effective argument to change the Skeptic's mind. We cannot rationally convince the Skeptic.

Does this mean that the Skeptic is right? Certainly not. It does mean, however, that we should stop trying to come up with arguments that will rationally convince the Skeptic to give up his skepticism. That exercise is a losing game. It has been rigged—by denying us all of the tools of rational argumentation—in such a way that we cannot possibly win . The fact that we are deprived of all means of rationally convincing the Skeptic shows only that we should not play this game. It doesn't show anything at all about the possibility of knowledge.

If someone asked us to show how it is possible to earn money without either working or investing, we might well be at a loss. Those are, it seems, the two ways in which money can be earned. But no one should conclude from this that it is impossible to earn money. Similarly, the fact that we cannot provide a dialectically effective argument against the Skeptic should not have us conclude that it is impossible to have knowledge.

I don't mean to suggest that we may simply rest content in our belief that knowledge is, in fact, possible and that this view requires no sort of defense at all. I don't believe that. There is work to be done, I believe, in showing just how knowledge is possible in the face of the skeptical argument. And I hope to rise to that challenge in the chapters which follow. But in showing how it is that knowledge is genuinely possible, we should not allow the Skeptic to dictate the rules of rational engagement. There is, I hope to show, a way to understand the challenge which makes sense of it and which doesn't turn the challenge into a pointless exercise. Showing how knowledge is possible should teach us something about the nature of knowledge, and trying to convince the Skeptic in a dialectically effective manner doesn't do that.

1.4. Reflective and Unreflective Belief Acquisition

There is a second important moral I wish to draw about the threat of skepticism, and it concerns reflective and unreflective belief acquisition.

Most of our beliefs are acquired unreflectively. I drive to work; I park my car; I walk across campus; and I settle into my office. In the course of performing these routine activities, my mind may be focused on many things. I may be thinking about my seminar later in the day and the issues I wish to discuss there. Or I may be thinking about the current political situation and how it is likely to unfold. I may be thinking about what I will prepare for dinner at the end of the day. But even if my mind is focused on these sorts of things, I will inevitably be picking up information about my environment as I proceed on my way from home to work. I could not possibly succeed in driving to work, rather than careening off the road, if I did not recognize where the road stops and the sidewalk begins. My ability to arrive at work each day is dependent on knowing where the road is and where the sidewalk is, whether the traffic light is red or green, whether there is someone crossing the street or not, and so on. I could not possibly park my car if I did not know where my parking lot is, and which spaces in it are occupied and which are free. I must know where the steering wheel is in my car and where the brake is found. I've been driving a very long time, so I don't need to focus on such things. But if I didn't know them, I couldn't possibly get the car parked. And so on. Just driving to work and walking across campus involves my forming an extraordinary number of beliefs as a result of the operation of my perceptual faculties even, as is typically the case, when my attention is focused elsewhere. The vast majority of my belief formation is unreflective. It just occurs within me without requiring my supervision, and I am not unusual in this respect. This is true of all of us.

We do, however, stop at times to reflect on our beliefs. We sometimes wonder whether beliefs we have are really ones we ought to have. I believe a colleague of mine is sure to turn down a job offer she has just received, but then I stop to think about this further and I wonder if this is just wishful thinking on my part. Do I really have good reason to think that she will stay here rather than accept what I know full well is an attractive offer? I believe that I've been spending too much money lately, and I should be saving more for my retirement, but then I stop to think about this further and I wonder if this is just overcautiousness on my part. Do I really have good reason to believe that I've been careless about my future? We all stop to reflect on beliefs we have in this way, at times, even if the overwhelming majority of our beliefs are not only formed unreflectively, but never reflected upon.

I not only reflect, at times, on beliefs I already have. I sometimes reflect in order to figure out what belief I should take on. I'm at a restaurant, and the check arrives, and I need to provide a tip for the waitstaff. At times, I just look at the bill and I instantly know, without having to think about it, what an appropriate tip is. But at times, I may be a bit distracted, or the bill may not be such as to lend itself to such instantaneous determination, and then I have to stop and reflect on what an appropriate tip might be. I engage in a bit of mental calculation, reflectively forming a belief when, in other circumstances, such reflection would not be required. Or a student asks me for advice about a situation I haven't confronted before. A response to her concern does not immediately spring to mind. I need to reflect on the issue she has raised in order to form a belief about what might be the best course of action. We all do this at times, reflecting on what to believe when belief does not just occur in us spontaneously.

There is a striking fact about the way in which the skeptical problem is typically posed. First it is pointed out that much of our belief acquisition is entirely unreflective, and then the possibility of error is brought to our attention. The source of error may be entirely unexciting: we sometimes overlook things, for example, when we fail to notice that we have left our car keys in the bedroom. Or the source of error may involve one of the nightmare scenarios detailed in section 1.2: we may be hallucinating, or dreaming, or deceived by an evil demon, or by a mad neuroscientist. Once we make note of these error possibilities, the fact that we typically form beliefs unreflectively, and never look back to reflectively examine whether we should have formed the beliefs we did, starts to look irresponsible. It is as if we are simply crossing our fingers and trusting to luck that our beliefs have been formed in the right sort of way, accurately registering features of our environment, rather than by way of bias, or misperception, or visual illusion, or worse.

This is exactly how Descartes proceeds, in what is surely one of the most striking and influential presentations of the skeptical problem, *Meditations on First Philosophy*. Descartes points out that he now realizes how many of the beliefs he formed early on in life turned out to be false, and that these misconceptions, no doubt, played a role, long before he realized his errors, in shaping his subsequent beliefs. His unreflectively formed beliefs, therefore, must be filled with error. He then points out that we are liable to simple perceptual error of all sorts, and finally, he brings out the big guns: the dreaming problem and the evil demon—the nightmare scenarios. Our unreflectively formed beliefs are shot through with potential for error, and we not only failed to consider these error possibilities when our beliefs were first formed, but we failed to reflectively examine these beliefs afterward, leaving them to infect our subsequent belief acquisition. There seems no reason whatsoever to put our trust in beliefs which are formed so casually and irresponsibly. Skepticism looms.

Descartes paints a powerful and horrifying picture of our epistemic situation, but the manner in which he raises the skeptical worry is hardly unique to him. Indeed, this way of proceeding has been quite typical in epistemology ever since Descartes. To take just one other example, the contemporary philosopher Laurence BonJour raises the very same issues in terms of what he calls *epistemic responsibility*. Simply forming beliefs unreflectively, as we tend to do, is just irresponsible, as BonJour argues at some length. A responsible person would not just sit back and allow their beliefs to be formed by happenstance, like a balloon tossed about in whatever direction the wind might blow. Responsible people don't live their lives in this way, acting thoughtlessly, leaving their behavior, and their character, to chance. By the same token, our belief acquisition should not be left to chance, or a mere hope for good luck, while our attention is occupied elsewhere.

The tradition thus paints a picture of unreflective belief acquisition as fraught with the possibility of error, something which, if left unsupervised by reflective thought, is bound to lead us astray. Knowledge is not the product of such irresponsible and chancy behavior.

If this is the problem, then the solution seems straightforward. As I mentioned earlier, Descartes is no skeptic, and neither is BonJour, nor is the philosophical tradition. Philosophers endorsing skepticism, as opposed to raising the threat of skepticism, are a rarity. If unreflective thought is what raises the skeptical problem, leaving the outcome of our belief acquisition to chance, the solution is to take our belief acquisition in hand, to be the masters of our fate, and to form our beliefs reflectively.[3] Reflective belief acquisition is not only required if we are to be responsible about our cognitive lives; reflective belief acquisition holds the key to explaining how knowledge is possible.

There is a great deal of disagreement among epistemologists as to how reflectively directed belief ought to proceed in order to secure knowledge. Descartes thought that we need to reflectively endorse belief in a god, by way of an argument which he provides, if we are to be in a position to have knowledge of the world around us. This aspect of Descartes's approach has not had many followers. It is not that most epistemologists are atheists or agnostics. Rather, our knowledge of ordinary things, such as my knowledge that I am now looking at my coffee cup, does not seem to depend on prior knowledge of, or belief in, a god's existence. But even if most philosophers have not followed Descartes on this particular point, the idea that the solution to the skeptical problem lies in reflective belief acquisition and reflective self-examination is a very common theme in the philosophical tradition.

The details, of course, matter, and there is a lot of interesting and important epistemological discussion about how reflective self-examination ought to proceed if knowledge is to ensue. But the issue I wish to bring attention to here is independent of those details. What is striking about the way so much of the philosophical tradition proceeds here is the lack of evenhandedness in the treatment of reflective and unreflective belief acquisition.

Unreflective belief acquisition is treated with a kind of default suspicion: it is, as it were, guilty until proven innocent. A parade of epistemic horrors is produced—misperception, bias, hallucination, and so on—and one is led to the conclusion that unreflective belief should not be trusted until one can prove it trustworthy. Without special reason to take our unreflective beliefs at face value, we should treat them as unjustified.

Reflection, on the other hand, is treated in exactly the opposite way. Notice that there is a very interesting empirical question about what the effect of reflective self-examination might be. When we stop to reflect on our existing beliefs to see whether they meet appropriate standards—whether we hold these beliefs for good-enough reasons— is the result of this self-checking procedure more accurate belief, less accurate belief, or is there no difference at all? This may seem like a silly question. The reason we engage in reflective self-checking is precisely to ensure the accuracy of our beliefs, so, of course, reflective self-checking makes us more accurate, or so it might seem. But while it is true that our motive for engaging in reflective self-examination is to improve our epistemic position, it doesn't at all follow from this fact that we are successful in achieving our goal.

Consider a simple case. When I tie my shoes in the morning, I do it unreflectively. I don't reflect on just what procedure needs to be followed—left over right, this lace under that, and so on—in order to succeed in getting my shoes properly tied. It is surely much the same for any adult, and for anyone other than a young child who is just beginning to learn how shoes are to be tied. Now ask yourself: would you be more successful, less successful, or equally successful in getting your shoes properly tied if you self-consciously thought about the proper procedure for tying your shoes as you tied them, being sure to follow that procedure as you thought things through? For most of us, this would interfere in a very big way with the smooth performance of the task. (Try it!) Here, reflective self-examination leads to less successful performance, at least for most of us.

The question I want to ask about belief acquisition and reflective self-examination is thus not a trivial one, and it is not obvious what the answer is. It might be that reflective self-examination makes us more accurate believers, or it might make us less accurate, or it might have little effect. It might produce better results in certain situations, and worse results in others, and make little difference in still others. We will return to this question, and look at the results of experimental investigation of reflective self-examination, in Chapter 4. For now, however, all I want to point out is that the question about the effect of reflective self-examination deserves to be asked, and that it is not obvious what the answer is.

It should now be obvious why I say that the philosophical tradition does not treat unreflective and reflective belief evenhandedly. What seemed to be an obvious thought—that unreflective belief can go wrong in all sorts of ways and so it is not to be trusted until it is reflectively examined—turns out not to be obvious at all. And, of course, it is worse than that. The seemingly obvious thought turns out to apply radically different standards to the two sorts of belief: unreflective belief acquisition is treated with default suspicion—guilty until proven innocent—while reflective belief acquisition is treated in just the opposite way—innocent until proven guilty. A fair-minded epistemological investigation cannot proceed in this way.

There are important questions to be asked about the reliability of the various processes by way of which we arrive at and revise our beliefs. But we need to ask these questions about both unreflective belief and reflective belief, and we need to do this in a way which does not put our thumb on the scale, prejudicing our investigation from the very beginning.

1.5. A Natural Starting Place

I want to make one final point about the way in which the threat of skepticism is typically treated. The skeptical problematic gets started by making vivid, as Descartes and so many others do, the possibility of error. Knowledge is first made to appear to be difficult: we must somehow avoid each of these error possibilities, possibilities which lurk around every corner. And then, as if that were not bad enough, we are confronted with arguments which make knowledge appear to be impossible. When we begin to think about the nature of knowledge in this way, we end up focusing our attention on a variety of situations—both the ordinary cases of familiar mistakes to which we are all susceptible, and the various nightmare scenarios which, of course, are merely imaginary. Our attention is focused in this way on situations in which knowledge is nowhere to be found.

Notice how odd it is to begin an investigation of the nature of knowledge in this way. If I want to understand the nature of living organisms, as biologists do, it would obviously be a very bad idea to begin by focusing my attention on environments in which living organisms are not even present. And if I want to understand the nature of social groups, as sociologists do, it would be a very bad idea to begin by focusing my attention on environments in which human beings live solitary lives, never interacting with one another. If you want to understand the nature of a certain phenomenon, you need to examine that very phenomenon, not look at situations where the phenomenon is entirely absent, let alone at situations in which that phenomenon could not even arise. No biologists or sociologists make this mistake, of course. The point about where to begin such investigations is too obvious to need to be stated.

But in the case of epistemology, the study of knowledge, this point needs to be stated, and it needs to be emphasized as well, because, since epistemology has for so long been motivated by the threat of skepticism and the peculiar situations which motivate skeptical worries, a vast swath of the epistemological literature has been devoted to understanding the nature of knowledge by focusing on situations in which knowledge is nowhere to be found.

So my suggestion, obvious as it may seem, is that we should begin our study of the nature of knowledge by examining the phenomenon of knowledge, rather than looking where knowledge is absent or impossible. And before we can theorize about the nature of knowledge, we need to provide some rough-and-ready survey of the phenomenon we wish to theorize about. That survey will occupy us in Chapter 2.

1.6. Conclusion

We are motivated to develop a theory of knowledge because of the threat of skepticism. Challenging arguments seem to show that knowledge is impossible, and an account of the nature of knowledge is needed in order to respond to these arguments. That is what a theory of knowledge would provide.

It is important, however, not to be misled by these arguments. We should not think that an adequate response to skepticism requires providing an argument which would convince someone who is himself a skeptic. A skeptic holds no beliefs and accepts no inferences as legitimate. Any argument we might try to offer the Skeptic, however, will require both premises and inferences from those premises. The Skeptic will thus regard any argument we might offer as a nonstarter. Argumentation cannot be dialectically effective against someone who will not accept any possible starting point for an argument and reject any possible transitions which argumentation might involve. But this doesn't show us that skepticism is correct. It shows, instead, that explaining how knowledge is possible does not require providing an argument which will rationally convince the Skeptic.

We have seen, as well, that the way in which the skeptical problem is typically presented treats unreflectively formed belief and reflectively formed belief in a way which is not evenhanded. Unreflectively formed belief is treated with deep suspicion: it is treated as guilty until proven innocent. We should not take such belief at face value, it seems, until we can provide special reason for thinking that it does not lead us astray. But then the solution to this concern, many suggest, is found in careful reflection on what we ought to believe, and such an approach abandons this default suspiciousness applied to unreflectively formed belief for the case of beliefs which are the product of careful reflection. Reflectively formed beliefs are treated as innocent until proven guilty. But we should not follow tradition here. Beliefs reflectively formed may be subject to error as well. In thinking about the ways in which our belief acquisition may lead us to error, we should not overestimate the likelihood of error when beliefs are unreflectively acquired, nor should we underestimate the likelihood of error when beliefs are the product of reflection.

Finally, we have seen that the way in which the skeptical problem is typically presented has us focus our attention on situations in which knowledge is impossible. But if we are trying to understand the nature of knowledge, this is a singularly inappropriate place to begin our investigation. If we want to provide an account of what knowledge is, we need to examine situations in which knowledge is actually found, rather than situations in which knowledge is not only absent, but in which the very features of the situation make knowledge impossible.

Suggestions for Further Reading

By far the best place to begin in understanding the skeptical problem, and, indeed, in thinking about epistemology, is René Descartes, *Meditations on First Philosophy*, 3rd edition, Hackett Publishing Company, 1993. For those with an interest in ancient skepticism, a good place to start is Sextus Empiricus, *Outlines of Pyrrhonism*, Loeb Classical Library, Harvard University Press, 1967.

Laurence BonJour's important discussion of epistemic responsibility may be found in his *The Structure of Empirical Knowledge*, Harvard University Press, 1985. BonJour provides an excellent introductory survey of epistemological issues in *Epistemology: Classic Problems and Contemporary Solutions*, 2nd edition, Rowman and Littlefield, 2010. For advanced readers, John Greco provides a very useful discussion of skepticism in *Putting Skeptics in Their Place: The Nature of Skeptical Arguments and Their Role in Philosophical Inquiry*, Cambridge University Press, 2000. Penelope Maddy provides an approach to these issues quite different from the one presented here in *What Do Philosophers Do? Skepticism and the Practice of Philosophy*, Oxford University Press, 2017. A strikingly original approach to the dream problem may be found in Ernest Sosa's *A Virtue Epistemology: Apt Belief and Reflective Knowledge*, volume 1, Oxford University Press, 2007. Brian Frances offers a defense of skepticism quite different from the more familiar Cartesian approach in his *Skepticism Comes Alive*, Oxford University Press, 2005.

2

The Phenomenon of Knowledge

2.1. The Beginning of an Inquiry

How should our inquiry begin? My suggestion is that we begin in just
the way other intellectual inquiries begin. We are trying to understand the nature of knowledge. We wish to develop some theoretical
understanding of what knowledge is, and we hope that our theoretical understanding will allow us to explain how knowledge is possible.
Since we don't yet have a theoretical understanding of what knowledge
is, we need to begin with our common-sense, pretheoretical understanding of it.

Consider an example from outside philosophy. Before there was
any theoretical understanding of the nature of water, before, in particular, it was known that water is H_2O, virtually everyone was very
well capable of recognizing water. They didn't, of course, recognize it
by way of its chemical makeup, since they didn't know what its chemical makeup was. Instead, they were able to recognize it in virtue of a
variety of superficial properties—it's a clear, colorless, tasteless liquid
at room temperature; it's the stuff that comes out of wells; it's the stuff
that comes out of the sky as rain and is found in lakes and rivers and
oceans. This recognitional capacity was reliable, but not perfectly reliable. When someone formed the belief that some liquid was water, they
were right much more often than not. But they were not always right;
mistakes were made. The fact that people were, for the most part, able
to recognize water when they saw it was enough to allow those who
wished to gain a theoretical understanding of what water is to begin
their inquiry. They could gather samples of water and try to figure out
just what it is that the samples have in common. As it turns out, this
was no easy task. But the inquiry could not even have gotten started
without a rough-and-ready recognitional capacity for water, the target
of the theoretical inquiry.[1]

Scientific Epistemology. Hilary Kornblith, Oxford University Press. © Oxford University Press 2021.
DOI: 10.1093/oso/9780197609552.003.0002

It's worth pointing out that this recognitional capacity need not work, and typically does not work, in any very self-conscious way. It's not as if young children have a mental list of various features by way of which they recognize water, self-consciously going down the list to check for relevant features before forming the belief that a particular liquid in front of them is, indeed, water. Most children would not be terribly accurate in describing how it is that they are able to recognize something as water. They might, to be sure, list a few common features of water if pressed to do so, but their recognitional capacity works unreflectively, and a self-conscious understanding of just how it works is not necessary for it to work, nor is it something which the typical child has. Nor is it something which the typical adult has. We too do not recognize liquids as water by self-consciously going down some mental checklist, and precisely which features of various liquids are the source of our classificatory judgments is not something which most of us know. We don't need to know such things in order to have the recognitional capacity.

We can tell a similar story about understanding the nature of gold. Before there was any theoretical understanding of the nature of gold— that it is, for example, the stuff with atomic number 79—many people had a rough-and-ready recognitional capacity for the stuff. Gold was a substance of great interest in many cultures long before there was any understanding of its chemical nature. It captured people's attention because of some superficial properties (its color, its brilliance, its capacity to be hammered into attractive and useful shapes), and although mistakes were certainly made in identifying some substances as gold which were not, in fact, gold (think of so-called fool's gold) and in failing to identify some substances as gold which were, in fact, gold, many people did have a tolerably reliable recognitional capacity for the stuff. And this was adequate to allow people to gather many samples of gold and begin a theoretical inquiry as to what it is, beyond the superficial qualities which initially brought the stuff to people's attention, that those of the samples that truly were gold had in common. Various acids could be poured over it to see which ones it was soluble in and which not. Its hardness and its malleability could be tested. And so on. As these tests proceeded, some of the mistaken identifications could be rectified. Once one sees, for example, that all but one of one's putative samples of gold are not soluble in hydrochloric acid, the one which does dissolve in hydrochloric acid probably isn't really gold. The capacity to separate samples of gold from those which are not gold improves as the inquiry proceeds.

Now water and gold are what they are in virtue of their chemical makeup, and knowledge, of course, is not a chemical substance. All the same, we can begin a theoretical inquiry into the nature of knowledge in much the same way that the inquiry into the nature of water and gold proceed. In particular, we begin by relying on our rough-and-ready recognitional capacity for cases of knowledge so that we may gather together instances of knowledge for further investigation. When someone buys a lottery ticket and believes that this time they will win that million-dollar jackpot, that belief is not a case of knowledge. It's not a case of knowledge, of course, when it turns out that the person doesn't win the jackpot. But even in that rare case where the person turns out to be right, and they do, in fact, end up winning, we can all agree that this is not a case of knowledge. They didn't know, even if they did believe that they would win. We all recognize that this is not a case of knowledge, even before we can say exactly what knowledge is.

Just as we recognize that false beliefs are not knowledge, and even certain cases of true belief are not knowledge, we often know full well that people know various things. If you wonder whether I know what time it is, and then you see me looking straight at my watch, then it is completely clear that I do, in fact, know what time it is. If Mary wonders whether Jack knows where her office is, and then Jack walks in the door of her office, Mary will recognize that, clearly enough, Jack does know where her office is.

Genuine cases of knowledge abound. I know where I parked my car this morning. You know what your name is. Most schoolchildren know that the earth is, to a first approximation, round, and that two and two make four. We achieve an extraordinary amount of knowledge while going about our daily lives without having to give matters a second thought. There is also knowledge which we can achieve only with a great deal of intellectual effort and careful investigation. Some knowledge is easily attainable by just about anyone, and some knowledge requires special expertise and training to achieve. As a result of the advancement of science, we know far more now than was known fifty years ago, or one hundred years ago, let alone a thousand years ago. These common-sense observations about knowledge, and many others like them, are enough to get our inquiry into the nature of knowledge started.

2.2. Children and Nonhuman Animals

Human adults are not the only creatures who deserve to be credited with knowledge. Children and nonhuman animals have a great deal of knowledge as well. (From now on, I will typically drop the qualifier and simply speak of "animals" when I mean "nonhuman animals.")

Preschoolers know where their toys are, often enough. They know that they like certain foods and dislike others. They know what they're wearing and whom they live with. We could not possibly explain their successful behavior—that they go to their toy box to get their stuffed animals rather than to the kitchen; that they reach to their right, where their food is, and not to the left, where it isn't, when they are hungry; that they run toward their parents rather than away from them when they want comfort; and on, and on, and on—without supposing that they have a vast storehouse of knowledge about themselves and their environments. Knowledge is what explains the possibility of regularly successful behavior, and even very young children regularly engage in a wide range of successful behaviors.

Although much of preschoolers' knowledge is revealed to us in what they say, these everyday examples show that even very simple behaviors, behaviors which hardly attract our notice, may be equally revealing of their knowledge. And once we see that regularly successful behavior can only be explained by attributing knowledge to these children, we see as well that even prelinguistic infants have a great deal of knowledge. One need not be able to speak or comprehend a language in order to take in information about one's environment and to act on that environment in ways which regularly succeed in getting one what one wants. Prelinguistic infants clearly recognize their parents and distinguish them from strangers: they know that this person is a stranger and that person is not. Once they start to crawl, they easily avoid obstacles in their path and move in the direction of things they desire. They reach out for various objects and succeed, in various degrees, in grasping them, thereby demonstrating their knowledge of the location of a multitude of things in their environment and what those objects are. Their knowledge of their environment increases by the day, and the sophistication of their exploratory behavior, which is instrumental in bringing about that knowledge, also grows by leaps and bounds. All of this knowledge is gained without a thought about the nature of knowledge, what counts as adequate evidence for reasonable belief, whether this belief or that is better supported by evidence, or whether they are subject to bias, deception, or misleading evidence. The conceptual capacities necessary for such sophisticated thoughts are years beyond these infants, but this does not prevent them from gaining knowledge every waking moment of their young lives.

The same is true of many nonhuman animals. Exactly where one should draw the line between those animals capable of knowledge and those, such as a variety of primitive microorganisms, which are not, is a complicated issue. But the line between knowers and living organisms who are entirely lacking in cognitive lives does not divide humans from all other animals. Dogs and cats are, without doubt, among the knowers. They know where their food is. Even when they cannot see their food dish, they come running for their food when they hear it being poured into their bowl because they know, on the basis of the characteristic sound the food makes on hitting the dish, that food is now in the bowl. They know where many objects in the household are to be found, and they display sophisticated problem-solving abilities, at times to the consternation of their owners, which serve as a rich source of newfound knowledge.

Animals in the wild demonstrate sophisticated cognitive abilities. They navigate their environment with ease in ways which indicate detailed knowledge. The hunting behavior of many animals, both individually and in social groups, is a product of sophisticated cognitive skills, and their responsiveness to changes in their environment and in the behavior of their prey make it clear, if it were not already, that many features of these behaviors are not mere reflexes, but a product of real learning and the ability to bring background knowledge to bear on their situation. The same is true in the behavior of animals avoiding

predators, and in their feeding and mating behavior. All of these sophisticated behaviors are made possible by the substantial knowledge-
gathering skills which these animals possess. Knowledge is a pervasive
feature of the animal world.

It should go without saying that these animals are not thinking
about epistemological issues, any more than human infants are. These
animals do not stop to consider whether they have adequate evidence
for their beliefs, or whether their beliefs might be a product of illusion, bias, or lack of sufficient care. Such sophisticated thoughts are
beyond them; they lack the conceptual repertoire necessary for enter-
taining such issues. Their lack of reflection on epistemological issues
does not, however, rob them of knowledge. They are robust knowers
nonetheless.

2.3. Adult Humans: Unreflective Knowledge

Most epistemological theorizing focuses on adult human knowledge, but I have started this chapter with less sophisticated knowers
for a reason. Adult human knowledge is a very special case. We adult
humans are far more sophisticated than other knowers in our conceptual capacities, in our capacity for reflection on our own mental states
and activities, in the background of experience and knowledge we may
bring to bear on our cognitive lives, and in the ways in which our cognitive activity is socially organized. It would distort our understanding
of the phenomenon of knowledge to focus our attention on such a
special case. As I have been emphasizing, even far less sophisticated
creatures than adult human beings are capable of knowledge.

And it bears repeating that most of adult human knowledge as well
is unreflective. Like other animals, and like our much younger selves,
we pick up an extraordinary amount of information without having to
stop to reflect on knowledge acquisition. A great deal has been written
in recent years about epistemic agency, the idea that knowledge and
justified belief may be the product of activities we undertake, rather
than things that merely happen to us.[2] If I attempt to solve some mathematical problem because the answer does not immediately spring to
mind, there are a host of things I do, mental activities which I engage

in, and not just processes that go on within me independent of my will. I may purposely think about similar problems which I have been able to solve, and focus my attention on various respects in which the problem at hand is either similar or different. I may think about various problem-solving strategies I have successfully employed in the past to see whether they can profitably be used to address the current problem. Lacking any more promising strategies, I may more or less randomly guess as to what the answer might be, and then employ some checking strategy to see whether my guess is correct, and, if not, whether it seems hopelessly wide of the mark, or, perhaps, if not exactly right, at least somewhere in the ballpark of a correct solution to the problem. All of these approaches to the problem involve things I do, rather than things that merely happen to me, and they are activities with a distinctive epistemic goal: I do these things in the hope of gaining knowledge. It is for this reason that epistemologists speak of epistemic agency: I am an agent in these situations, doing things to bring about knowledge, and not just the locus of various processes occurring within me. We will say a bit more in the next section of this chapter about such epistemic activity, and still more in Chapter 4, but for all the importance of epistemic agency in a survey of our pretheoretical view of the phenomenon of knowledge, it is crucial that we not give the passive reception of information short shrift, even in the case of adult human beings.[3]

We need not be agents, we need not do anything, we need not focus our attention or think of problem-solving strategies, in order to gain knowledge. As long as we are awake, there are numerous processes which will go on within us producing knowledge. Just as we need not do anything in order for our hearts to pump blood through our bodies, we are possessed of cognitive mechanisms which go to work within us without any need for our attention or will. We are constantly acquiring knowledge throughout our lives just as passively as we acquire gray hair as we age.

First and foremost among these cognitive mechanisms are our perceptual systems. Perception is not exclusively a passive affair: we sometimes need to focus our attention, or turn our heads, or think through just what it is we are actually seeing, in order to gain perceptual knowledge. But these are the exceptional cases rather than the rule. Our perceptual systems work within us, cranking out knowledge of the world around us, while our attention lies elsewhere or nowhere at all. We could not manage to walk down a flight of stairs, or eat a meal, or carry on a conversation, or any of the other activities of our daily lives were this not so. Successfully carrying out these activities, and count-less others, is dependent on knowledge of innumerable details of our situation—the height of the steps on the stair, the distance from the bottom of the staircase, the location of the various foods on our plate, the orientation of our utensils, the impact of what we say on our con-versational partners. If we had to attend to each of these things and direct our mental activity in every case rather than merely pick up all this information passively, we'd never be able to make it out of bed in the morning. Beliefs are constantly being produced within us, without our notice, and without any need for our direction; our ability to act is deeply dependent on this passive reception of information.

It is not just straightforwardly perceptual beliefs which are pro-duced in us passively. Inferential consequences of our beliefs may be drawn out by our mental machinery just as passively. While there can be little doubt that inference is sometimes the product of self-conscious consideration and active engagement, as in the example of solving a mathematical problem where no answer immediately springs to mind, inference without such self-conscious consideration or active engagement is a regular feature of our cognitive lives. If I believe that I left my car keys on top of my dresser, and then go to retrieve them but find that they are not there, the belief that my keys are not on the dresser is not simply added to my stock of beliefs. If it were, I would now believe both that my keys are and are not on the dresser. Rather, the new belief replaces the old, without my having to think about the simple logical point that it would be a mistake to form contradictory beliefs. My cognitive machinery automatically has one belief replace the other without any need for active attention or supervision on my part. Similarly, in discovering that my keys are not on the dresser, I come to believe that I need to continue my search if I'm to be able to drive to work. This consequence of failing to find my keys on the dresser is not something that I have to self-consciously draw out, as in the case of the mathematical problem. My cognitive machinery au-tomatically draws this conclusion for me without my having to give it a second thought. Every time I take in new information perceptually, certain obvious consequences of that new information relevant to the tasks I'm engaged in are automatically produced within me. Here too, were this not so, it would be impossible to get through the day. Our ability to function successfully requires the smooth operation of a vast array of cognitive mechanisms operating within us without need for our attention. There are only so many things we can actively focus on at any given time, but our need for information to direct our activities greatly outstrips the capacity of attention. Our automatic cognitive mechanisms, constantly bringing in new information by way of perception, and drawing out certain obvious consequences of those perceptual beliefs relevant to our activities, make the activities of our daily lives possible.

2.4. Adult Humans: Reflective Knowledge

Even if the overwhelming majority of our belief acquisition and revision goes on without our having to reflect on our beliefs, there can be little doubt that we do, at times, stop to reflect. We sometimes think about existing beliefs of ours, wondering whether we really ought to believe as we do; and we sometimes think about what beliefs we ought to take on, even before our automatic mechanisms have provided us with some belief. This capacity to reflect on our beliefs seems to be unique in the animal world. At least as far as casual observation goes, there seems little evidence that other animals do, or even can, engage in reflective self-examination. We can't begin to explain why it is that dogs and cats go running to their food bowls when they hear food being poured into them without supposing that they have a good deal of knowledge about what is happening around them. But none of their behavior seems to call for the supposition that they stop to reflect on whether they have adequate evidence that there is food in their bowls. The supposition that they should engage in such reflection seems absurd. Nor is there any obvious evidence that they ever stop to reflect, in prospect, about what to believe in some challenging situation. Casual observation of other nonhuman species does not offer encouragement on this score about them either. I don't mean to suggest that casual observation is sufficient to address the question about animal cognition.

It isn't. But this chapter is addressed to our pretheoretical views about knowledge, and at least pretheoretically, there seems no reason at all to attribute reflective belief acquisition and revision to any species other than our own.

Let us take the two kinds of reflective self-examination separately. We are sometimes challenged by others on some opinion we have offered. "Why do you think that?" someone may ask us. When we are asked to offer reasons for our beliefs in this way, we may, at times, come to wonder whether we really do have good reason to believe as we do. We stop to reflect on our reasons and find, at least at times, that we are brought up short: it's not clear why we believe as we do, or it's not clear that the reasons we can think of really do give us good reason for the belief we have. In the face of such a situation, we may come to abandon our belief. We come to think that we never did have good reason to form the belief we did, and we give it up.

Of course, things don't always turn out badly for us when we are challenged in this way. Often enough, we are easily able to meet the challenge. When asked why we believe as we do, we may promptly offer very substantial reasons in support of our belief. Such reasons may not only satisfy us; they may often satisfy our challenger as well. Reflective self-examination can give us reason to go on believing as we did, and even give us reason for greater confidence in our opinions. Once we succeed in making our reasons explicit, something we may not have needed to do before being challenged, we may find that our belief is perfectly in order as it stands.

Reflective self-examination of this sort is often prompted by challenge from others, but we may also engage in this activity without such prompting. On noticing that my neighbor's political beliefs seem to be fed on a one-sided diet of reading material, looking only at news sources that are likely to support what he already believes, I may turn reflective and wonder whether I am any better in this regard. This may lead me to think carefully about whether my own political opinions are well supported, even if I can offer impressive arguments in their favor, because it leads me to suspect that I might have been shielding myself from counterevidence. Our capacity to engage in such self-examination and self-criticism may prompt not only change in our beliefs, but changes in the ways we go about getting evidence.

We are prompted to reflect on what beliefs we might adopt, in prospect, most typically when we find a need for belief on some issue, and yet no belief is available to fill the need. Perhaps I'm driving home from some place that I've just been to for the first time. The route was unfamiliar to me, and I had complicated directions to follow on the way there. Now, on the return trip, I just assume I'll remember the route, but I find myself at an intersection unsure whether I should turn left or right. "Which direction did I come from?" I think to myself. Previous intersections prompted no such reflective thought because at every point thus far I simply knew, without having to think about it, which way to go. In thinking about what I ought to believe—whether the route home involves going left or right—I may carefully scrutinize features of the landscape, looking to see some familiar sight, hoping that such careful scrutiny will make the answer to my question clear.

Or I might engage in some home repair project, initially thinking that I know how to carry it out successfully. I need to install a small electric fan to circulate the air from my wood-burning stove, and I decide to wire it in to some preexisting circuit. As I start to attach the wires, I realize that I don't remember just how to do this. Do I need to remove the wires from the light switch and splice in the new wiring to the fan, or do I leave the wiring to the light switch in place and just connect the new wires on top of the old ones? It occurs to me that there's something I learned long ago about parallel and series circuits, and for the life of me I can't bring it to mind.

Reflective self-examination of this sort may lead us to recognize the shortcomings of our epistemic position, that we really don't know what we ought to believe, and that we need to gather additional evidence: look back at the driving directions to see which way to turn, or consult some do-it-yourself book to see how the wiring is properly done. But we sometimes find that further evidence gathering is not really necessary. We have all the evidence we need; we just need to figure out what belief our evidence supports.

I've been thinking about some economic policy which is currently being debated in Congress. The policy is controversial, and the consequences of implementing it, both good and bad, have been much in the news. I've followed the back-and-forth about these advantages and disadvantages, but I haven't yet reached an opinion about whether this is a good policy or not. I can reliably recite the evidence: the good and bad consequences of the proposed policy. I know how these consequences compare to various other policies which have been proposed. But I haven't put it all together; I haven't figured out, all things considered, whether the policy currently being discussed is a good one.

On less complicated issues, I don't need to stop to think things through. Once I see the evidence, I am immediately convinced; I know what to think. But this issue is complicated enough that I have to think through the evidence; I need to stop to reflect on what the evidence shows.

Such situations are not altogether rare. They are an important part of our cognitive lives, and, at least on its face, it seems that some of our greatest epistemic achievements are a product of such self-conscious reflective thought. Even if much of our knowledge is easily attained, indeed, so easily attained that it requires no effort on our part at all, there is a good deal of knowledge as well which is a hard-won achievement, requiring careful reflection about just what we ought to believe. Any adequate survey of the phenomenon of knowledge must encompass both sorts of knowledge.

2.5. Social Dimensions of Knowledge

I have been speaking thus far, for the most part, as if knowledge is an individual achievement. Perceptual knowledge is typically a solitary affair. The coffee cup is in front of me; I see it and I thereby know it is there. Even if there is someone else in the room with me, they have nothing at all to do with my gaining perceptual knowledge, at least in the typical case. Much of our inferential knowledge seems equally solitary. Often enough, I need no input from anyone else to draw out the consequences of my beliefs.

But humans are social animals, and there are important social dimensions to knowledge acquisition. We have already indicated that some of the stimulus to reflect on our beliefs may come from others: they ask us why we believe what we do; they challenge the reasons we offer; they present us with counterarguments. Such social interactions play a tremendously important role in the acquisition of knowledge.

And even when no other person is physically present, much of my knowledge is acquired in a way which is clearly dependent on others. I read books, and newspapers, and magazine articles, and websites. Without such input, the scope of my knowledge would be vastly diminished. I've never been to Buenos Aires, or Perth, or Hong Kong, but I know a good deal about them. I've never done research in physics, or chemistry, or biology, but I'm not entirely ignorant about these subjects. In every case, my knowledge of these matters is indirect, depending on the say-so of others whose knowledge is, in many cases, far more direct than mine.[4]

Our knowledge is dependent on others not only on abstruse matters, but even on some of the most mundane matters we can imagine. I know what my name is. For each of us, our knowledge of our own name is about as certain as any knowledge can be. With the exception of those who are suffering from a very extreme and rare form of amnesia, knowledge of one's own name is second nature. We never give it another thought, nor need we, and no one would ever challenge us on such a matter, suggesting that, perhaps, we don't really know what our name is. But what is the basis for such knowledge? It is not straightforwardly perceptual, like my knowledge of the presence of my coffee cup. I know what I look like by looking in a mirror, but that doesn't help with knowing my own name. People address me by my name, and I see it listed outside my office and on my driver's license, and credit cards, and various pieces of mail that arrive at my house. We don't typically think of these things as any part of the basis for our knowledge about what our name is, probably in part because we don't typically think at all about what the basis for that knowledge is, but we can see how these things are importantly relevant by imagining what our reaction would be if people started addressing us with some other name, or if we found that the documents in our wallet all had some other name on them, even if they had our photograph. The fact that such shocking things don't happen to us shows that we are constantly receiving additional evidence about what our name is, even if the matter is so firmly established already that no additional evidence is needed. All of these sources of evidence are social: they are dependent on the knowledge of others. The same is true, of course, of the knowledge that my parents and siblings and friends, as far back as I can remember, called me by my name, and that I remember seeing it as well on my birth certificate, even if I can't remember when I last looked at that document. My knowledge of what my name is, even if it is at least as secure as my knowledge that I am looking at a coffee cup now, is far less direct, and it is thoroughly dependent on the knowledge of other people.

More than that, all of our knowledge is influenced and informed by the people who surround us from the moment we are born. It's not so much that our parents instruct us on various matters, though that surely is the case. Our ways of seeing the world, what we take for granted and what we see as standing in need of reasons, are themselves something that our upbringing communicates to us in subtle and not so subtle ways. It would be one thing if children only believed what their parents told them when the children had independent evidence of their parents' reliability on the matters in question. But, of course, nothing could be further from the case. Especially early on, children are in no position to check on the reliability of what their parents tell them, let alone on the subtle messages which are communicated to them in less direct ways. Our upbringing colors our view of the world from the very beginning of our lives, and even when we come to reject various parts of what we learned and came to think, we do so against the background of many features of that inherited worldview.

I don't mean to suggest anything diabolical about this fact, although there are certain situations which can result in children having large bodies of false beliefs. (Even in such epistemically unfortunate cases, this is nothing like Descartes's worry about an evil demon.) The typical case is very far from diabolical. Parents do, in most cases, have their children's interests at heart, and they succeed in communicating a wealth of information about the world in ways that inform, rather than distort, their children's bodies of beliefs. My point here is not that there is something problematic about inheriting, as it were, a good deal of what one believes from one's parents. I simply mean to point out that much of what children believe is secondhand knowledge, rather than knowledge which is gained in the direct manner involved in simple cases of perception.

Our beliefs, from our earliest days, are a product of our reliance on others. We should not view children as autonomous knowers. These social influences on our beliefs continue throughout our lives, even as we become more sophisticated in our thinking and more capable of independent evaluation of our beliefs. As adults, how we evaluate our beliefs is influenced by input from others, including how others respond to what we tell them, what they challenge us on, and what they allow to pass without notice. If we think about the ways in which our beliefs are acquired and revised by thinking about an individual knower operating alone and without any social sources of influence, we will not be thinking about knowledge as it is actually found in human beings.

2.6. Conclusion

Knowledge is found not only in human adults, but in children, even prelinguistic infants, and in many nonhuman animals as well. We can only explain the successful behaviors we see in others by supposing that they have a good deal of knowledge of features of their environment. Young children and various animals are able to gain this knowledge without the benefit of the capacity to reflect on their beliefs or their knowledge-gathering activities. They don't stop to reflect on whether they have adequate evidence, or whether their circumstances are somehow misleading. Their beliefs are produced in them by cognitive processes operating within them automatically and not requiring any self-conscious direction or supervision. The fact that such creatures do not give any thought to their knowledge-gathering activities or to the strength of the evidence they have for their beliefs does not rob them of knowledge. They have very substantial bodies of knowledge despite these cognitive limitations.

Human adults are more sophisticated knowers. We have the capacity to reflect on what we're doing, and we sometimes reflect on our beliefs. We wonder, at times, about whether the beliefs we have are ones we really ought to have, and we sometimes reflect on what beliefs we ought to adopt, giving careful thought to the strength of the evidence we already have and, at times, what evidence we need to gather if we are to form an opinion at all. But much as we are capable of such sophisticated cognitive activity, we too often arrive at our beliefs unreflectively. Indeed, this is the typical case. And beliefs arrived at unreflectively are a very substantial part of the knowledge we have.

A focus on simple cases of perceptual knowledge can give the misleading impression that knowledge is exclusively an individual achievement. Social factors, and reliance on the epistemic activities and achievements of others, are a pervasive feature of human knowledge. It is not just the sophisticated social organization of scientific research which illustrates our reliance on others in gaining knowledge. Human knowledge, from the very beginning of our lives, shows a deep dependence on the knowledge of others. We take on beliefs and ways of thinking about the world from those around us long before we are capable of any kind of independent check on their accuracy or reliability. Such checking is only possible against the background of a substantial body of beliefs about the world, a rich understanding of a sort which itself could not be achieved without prior reliance on the knowledge of others. Human beings are not autonomous knowers. The extraordinary range of human knowledge, unrivaled by other species, is a product of the ways in which our knowledge is a social, and not just an individual, achievement.

All of this is little more than our pretheoretical picture of the phenomenon of knowledge. We can, however, deepen our understanding of this phenomenon through more systematic investigation. We begin that investigation in the following chapter.

Suggestions for Further Reading

For a very useful introductory discussion of knowledge acquisition in children, see Alison Gopnik, Andrew Meltzoff, and Patricia Kuhl, *The Scientist in the Crib: Minds, Brains, and How Children Learn*, William Morrow, 1999. A philosophically rich discussion of knowledge in nonhuman animals may be found in Colin Allen and Marc Bekoff, *Species of Mind: The Philosophy and Biology of Cognitive Ethology*, MIT Press, 1997. A discussion of animal knowledge from an evolutionary perspective may be found in John Alcock, *Animal Behavior: An Evolutionary Approach*, 10th edition, Sinauer Associates, 2013. My own book, *Knowledge and Its Place in Nature*, Oxford University Press, 2002, discusses human knowledge as a special case of the broader phenomenon of animal knowledge. A focus on the social aspects of early cognition in humans may be found in Paul Harris, *Trusting What You're Told: How Children Learn from Others*, Harvard University Press, 2012. A broad discussion of the importance of social factors in human cognition, presented from an evolutionary perspective, may be found in Joseph Henrich, *The Secret of Our Success: How Culture Is Driving Human Evolution, Domesticating Our Species, and Making Us Smarter*, Princeton University Press, 2016. Alvin Goldman's *Knowledge in a Social World*, Oxford University Press, 1999, is a tremendously important philosophical discussion of the social dimensions of knowledge.

3

Knowledge from the Outside
The Third-Person Perspective

3.1. What the Third-Person Perspective Has to Offer

We may examine the phenomenon of knowledge from either of two perspectives. We may take a third-person perspective, the perspective of an external observer, and view knowers in the way that experimental psychologists do. Alternatively, we may take a first-person perspective on knowledge, the perspective which knowers themselves have on their own belief acquisition and revision. In this chapter, we take the third-person perspective and look at knowledge from the outside, as it were; in the next, we look at knowledge from the inside, the perspective knowers have on their own cognitive lives. There is a good deal of motivation for adopting each of these perspectives, and a theory of knowledge has much to gain by looking at knowledge in each of these ways.

The reasons for adopting the third-person perspective are, I believe, straightforward. As we have seen, much of our cognitive lives is a product of automatic processing, psychological processes which take place within us below the level of conscious attention. Much of the perceptual knowledge we acquire which makes our daily activities possible is of this sort. When I walk across campus thinking about what I'll talk about in my seminar meeting later in the day, my attention is focused on philosophical and pedagogical issues; it is not focused on features of the walkway, or the people on it, or the best route to take to get where I'm going. Nevertheless, the fact that I succeed in staying on the walkway without bumping into people, that I easily negotiate irregularities in the pavement and chart a course around various obstacles, that I get where I'm going, can only be explained by supposing that I pick up information about all these matters as I walk along. I come to

Scientific Epistemology. Hilary Kornblith, Oxford University Press. © Oxford University Press 2021.
DOI: 10.1093/oso/9780197609552.003.0003

know about all of these things even though I'm not focusing my attention on any of them. For that very reason, the first-person perspective on my knowledge of these matters is utterly useless. I don't have a first-person perspective on that knowledge; it's invisible to me. So if I want to understand the nature of the knowledge which automatic processes provide, a third-person perspective is required.

And this is no small matter. Because so much of our knowledge is a product of automatic processing, the third-person perspective promises to tell us a great deal about our knowledge that a first-person perspective misses out on. If we wish to understand the nature of knowledge in general, and not just that small corner of our knowledge which occupies our conscious attention, the third-person perspective is absolutely essential.

This would be reason enough for taking the third-person perspective on knowledge, but it is not the only reason. The perspective of the experimental psychologist is a scientific perspective, and it offers all the benefits that science has to offer. Well-tested and well-confirmed theories in science offer more than just one perspective among many. Casual observation, seat-of-the-pants guesswork, and the wild speculations of one's crazy uncle may offer perspectives on just about anything, including the nature of knowledge. But the third-person perspective we will be examining here, the third-person perspective offered by a scientific investigation of cognition, is not just one more perspective to be placed alongside these others and treated as their equal. Well-confirmed theories within the sciences offer our best current understanding of the phenomena they investigate. If we wish to understand, for example, how perception works, we need to turn to a scientific investigation of perceptual processes. It's not just that a first-person perspective on these processes offers no illumination. Rather, the third-person perspective offered by science on these matters is our best available route to understanding them.

It is impossible, in a book of this size, to discuss all of cognition. We can, however, discuss some features of the processes by which we gain knowledge which offer special insight into the question of how knowledge is possible. In this chapter, I will focus on perceptual knowledge and inference. There are general lessons to be learned here which not only tell us about human psychology, but tell us about the nature of knowledge itself.

3.2. Perception

Consider a simple case of perceptual knowledge. I'm sitting at my desk, and over the past few days, there's a book I've been looking for which I've been unable to find. You just located the book, and you walk into my office and gently toss it to me from across the room. The book flutters through the air straight to me and I catch it. Well before I catch it, I recognize that it is a book which you've launched into the air. I know that the object headed toward me is a book. This may seem to be no great cognitive achievement, but it is a case of knowledge nonetheless. My knowledge that the object is a book is a case of perceptual knowledge.

How was I able to gain this knowledge? The first-person perspective is of little use in illuminating this matter. If you were to ask me how I knew the object was a book even before I caught it, I'd probably say that I could just see that it was a book; after all, I was looking right at it. But that doesn't tell us much. Of course I saw it. But how was I able to do that?

Let me clarify the question I am asking here. Recognizing that the object tossed to me was a book is a far more complicated achievement than it at first seems. Indeed, even recognizing that the object tossed to me was a roughly rectangular solid involves some interesting psychological processing. Let me focus then on the latter bit of knowledge: my knowledge of the shape of the object I see.

Light strikes the book, bounces off the book, and then forms an image on my retina. (It actually forms an image on each of my two retinas, but I'll pretend that I have only a single eye so that the complexities of binocular vision can be avoided.)[1] As the book moves through space, the shape of the image on my retina changes continuously. Let us suppose that, at one point, the book is at eye level, the cover facing me, and exactly perpendicular to the floor. From such an angle, the book would cast a rectangular image on my retina. I would not be able to see its edge, or any part of the back of the book. Nothing but the front cover would be visible to me.

Now imagine that the bottom edge of the book moves slightly closer to me as the book turns in the air. Instead of casting a rectangular image on my retina, the book now casts an image which is trapezoidal in shape: the bottom edge of the book, being slightly closer to me than its top edge, casts a slightly larger image than the top of the book.

The book continues to turn as it moves through the air, and now it twists slightly so that, in addition to the bottom edge being closer than the top edge, the right side of the book is slightly closer to me than the left side. Now the image which the book casts on my retina is no longer trapezoidal. It has the shape of an irregular quadrilateral.

As the book continues toward my hands, it turns on its edge, with neither the cover nor the back of the book still visible. If we imagine that it is a very thin book, the image it now presents will be of a very thin rectangle, barely thicker than a line.

All of these different shapes cast on my retina are somehow processed by my visual system so that I recognize the object moving toward me as a rectangular solid, despite the fact that the vast majority of those images were not rectangular. Indeed, had the book never been at just the proper angle in facing me, it would never have cast a rectangular image on my retina at all, and yet my visual system would have worked just as well, easily allowing me to recognize the three-dimensional shape of the object almost instantaneously. The book passed through space in but a brief moment, and yet my visual system somehow processed that ever-changing parade of images in such a way that the shape of the object was never in doubt for me.

And now the question I want to ask becomes clearer: how was my visual system able to do *that*? What seemed initially to be an utterly trivial bit of knowledge acquisition—recognizing the shape of the object headed toward me—is now revealed as a remarkably complicated piece of information processing. All my visual system has as input is a series of ever-changing shapes in two spatial dimensions, and yet somehow it is able to determine the shape of the object projecting those images. What allows the visual system to extract this information from such a source?

If you think of the visual system as designed to work in any imaginable environment whatsoever, you will make this complicated information processing task even harder than it actually is. Indeed, you will make it impossibly hard. Here are some nightmare scenarios for the visual system. Perhaps you have read Edwin Abbott's *Flatland*, a fanciful story about a two-dimensional world, and you consider what it would be like for our visual system, working in the way it in fact does, to operate in such a world. Alternatively, one might imagine a world in which objects have four or more spatial dimensions. One might instead imagine a world in which objects have three spatial dimensions, just as our world does, but the objects constantly change their shapes in wild and unpredictable ways. One could even imagine a world in which objects pop in and out of existence, but when they go out of existence, they are immediately replaced by other objects of different shapes. All of these worlds and more are certainly imaginable. It should be clear that it is impossible to design a visual system which would work well in all of these different environments. If the objects the visual system needs to detect might have any of these different characteristics, the visual system would not have input rich enough to distinguish among the myriad possibilities. If the ever-changing pattern of images projected on the retina might be the product of a single object with relatively stable boundaries, or an object which is constantly changing its shape, or countless objects popping in and out of existence, each taking the place of another of different shape, no possible visual system would be able to reliably extract information about the shape of objects providing it with input.

What this shows is that our visual system is able to do the job it does because it is not even trying to discriminate among all of the imaginably different ways the world could have been. Instead, the visual system, in effect, takes certain things for granted about the world in which it operates. It takes for granted, as default assumptions, that the objects projecting images on the retina have three spatial dimensions; that they persist through time, rather than popping in and out of existence; and they have relatively stable boundaries, which are not constantly changing at random. Only by making these default assumptions is it possible for the visual system to do its job.

Let me explain what I mean by talking about the visual system "taking certain things for granted," or "making certain assumptions," or "presupposing" various things. The visual system is not, of course, a little man, or a little woman, inside our heads. It does not literally take anything for granted, or make any assumptions, or presuppose anything. But the visual system operates in such a way that, if certain things are typically true of the world around it, it will tend to produce accurate results. We may therefore usefully think of these things that would have to be true of typical environments in order for the visual system to work reliably as things that the visual system takes for granted, or assumes, or presupposes.

If an engineer were building a visual system, perhaps for a robot, the engineer might note that there are certain regularities in the environment in which the robot operates, and instead of trying to design a visual system for the robot which would figure out these regularities, the system would simply exploit these regularities in doing its job. The engineer might build information about these regularities into the memory of the visual system, so that the interpretive process the visual system engages in would always draw on this background information provided by the engineer. This is one way in which the system could work. But it need not work this way.

Consider the design of a screwdriver. It has a handle, designed for gripping by the human hand; a rigid shaft, which will not shatter when subjected to the stress of turning a screw; and an end which is shaped so as to fit squarely within one end of a typical screw so that, when the screwdriver is placed within the screw, turning the screwdriver will result in turning the screw. We may say that the design of the screwdriver presupposes that screws will not change shape as one attempts to turn them, or that they will not shatter under the pressure of rotation. Of course, the screwdriver does not literally presuppose anything. Nor is it the case that the screwdriver has some sort of built-in memory in which these facts are stored. Rather, it is these facts about screws— that they don't change shape as one tries to turn them and they don't shatter under the pressure of rotation—that allow the screwdriver, given its physical makeup, to operate successfully in the environments in which it is typically used. A well-designed screwdriver would not work well in just any imaginable environment. The best screwdriver you can buy would not work well in an environment in which screws constantly change their shape, or are made of glass and shatter under the slightest pressure. But we don't live in a world like that and excellent screwdrivers may take advantage of that fact.

The same is true of the visual system. Talk of what the visual system presupposes, or takes for granted, and so on, is just shorthand for saying that the way in which the visual system operates allows it to reliably extract information about environments in which certain regularities hold. If the visual system were designed, as visual systems for robots are, then these regularities might be presupposed or taken for granted by the engineers who designed it. In naturally occurring visual systems, however, such as those in humans and other animals, there need be no designer. We can recognize that these naturally occurring visual systems work well because they take advantage of regularities in the environments in which they operate. It is precisely by exploiting such regularities that the visual system can do its job.

We can figure out what these regularities are by trying to concoct situations in which we can fool it. This is why perceptual psychologists are so interested in visual illusions. By seeing where the visual system goes wrong, we can discover just what the regularities are in normal environments which allow it to work reliably so much of the time.

Consider, then, the phi phenomenon. Under certain conditions, the visual system can be tricked so that there seems to be a moving object even when nothing is moving at all. Road signs sometimes take advantage of this illusion. A highway diner may attract the attention of passersby with an illuminated sign in the shape of an arrow, pointing in the direction of the diner. The sign consists of a very large number of lightbulbs which are turned on sequentially, with the ones furthest from the diner illuminated first, followed by the next row of bulbs closer to the diner, and then the next, until the entire sign is lit. This creates an illusion of motion, as if the arrow is somehow beckoning travelers toward the diner. Signs such as this capture our attention. We can't help but notice them. Nothing is actually moving on the sign. There is only a sequence of bulbs turning on and off, but it looks for all the world as if the illuminated arrow is moving.

The same illusion occurs when movie marquees have a border of lightbulbs which are turned on and off in sequence. Every forth bulb, say, on the perimeter of the marquee is initially illuminated briefly, and as soon as it goes out, the bulb to its right is turned on for the same length of time. When it goes out, the bulb to its right is briefly illuminated, and the process continues. Nothing is moving on the sign; bulbs are just going on and off in sequence. We are presented, however, with the illusion of motion, as if the lights are circling rightward around the perimeter of the marquee. The illusion that they are moving in the opposite direction can be produced just by reversing the order in which the bulbs are illuminated.

The phi phenomenon is not only exploited outside the movie theater. Movies themselves trade on our susceptibility to this illusion. Movies consist of a series of projected still images. If the images were projected sequentially at a very slow rate, with only a few images projected each second, we would not have the impression of motion. For example, a series of such images of a woman walking across a room would not look like a woman walking at all. We would see her in one position, and then a second, and then a third, but we would not have the impression of motion. Speed up the rate at which the images are projected, however, and, eventually, we no longer see a series of still images; we have the illusion of motion. The visual system gives us the impression of a person moving despite the fact that the input to the visual system is nothing more than a series of still images.

In each of these cases—the road sign, the movie marquee, and the movie itself—we are presented with a series of images where the source of those images involves nothing in the environment which is moving. Despite that fact, we can't help but have the impression of movement. Even when we are aware of how the road sign works, or the movie marquee, or the movie itself, we still have the impression of motion. This is true of visual illusions generally: knowing that we are subject to such illusions does not make them go away. This tells us something very important about how the visual system works.

Our visual system presupposes, in the sense defined previously, that the world around us is typically populated by objects with three spatial dimensions and which generally maintain their shapes as they move through space. When a sequence of images is projected on the retina, the visual system acts so as to reconstruct what those three-dimensional objects must have been in order to project that sequence of images. As a result, when you toss the book to me from across the room, the visual system doesn't need to figure out that there is a three-dimensional object with a fairly stable shape moving toward me. By taking all of that for granted, it is quickly able to perform the otherwise impossible task of determining the shape of the moving object. Not having to figure out, among all of the imaginable objects that might be out there, including ones which pop in and out of existence, ones which are constantly changing shape, and ones which have two or eight or ten spatial dimensions, what shape or shapes are responsible for the images on my retina, an impossible task is thereby transformed into an admittedly difficult, but doable, task: figure out, under the assumption that the object out there is a three-dimensional thing which keeps its shape as it moves, just what the shape of the object is. Perceptual psychologists have, in fact, been able to figure out how the visual system performs this task, once the problem is constrained in this way.

Of course, when the visual system builds in such presuppositions about its environment, it will inevitably make mistakes in cases in which those presuppositions are false. And that is what happens in the case of the visual illusions. The sequence of lightbulbs going on and off in the road sign and the movie marquee present the sort of sequence of images on the retina that would be projected by a single object moving through space. Since the visual system takes for granted that it is trained on such objects, it provides us with the impression of a moving arrow on the road sign, or of something moving around the perimeter of the movie marquee. And when we watch a movie showing a woman walking across a room, even though the images in the sequence projected on the screen, as well as the images on our retinas, are all unmoving, we nevertheless have the unshakable impression of motion. The visual system, built to assume that it is focused on a world of three-dimensional objects which keep their shapes as they move, has figured out what shape such an object must have to have projected that sequence of images. The very feature of the visual system which allows it to work so well under normal conditions provides us with illusions when conditions are, in one way or another, out of the ordinary.

It is for this reason that studying visual illusions allows us to understand how the visual system works so well. When we see where the visual system can be made to go wrong, we thereby see what it is that the visual system is presupposing about typical environments. And it is by making such presuppositions that the information-processing task of identifying the source of our retinal images becomes tractable.

I have said that knowing we are subject to visual illusions does not make them go away. This does not mean, however, that we are always fooled by visual illusions or that they always result in false beliefs. Once you know about the phi phenomenon, you need not come to believe that there is some moving object that is the source of your visual experience. The impression of motion is unavoidable, but learning about the illusion allows you to avoid giving in to the inclination to believe what your experience inclines you to believe. The way in which the visual system works is unaffected by what you learn, but the visual system does not directly produce beliefs. It produces visual impressions, and inclinations to believe. What we learn can influence how those impressions, and those inclinations to believe, produce further effects downstream. The visual system is tamperproof; it is not affected by what we come to believe, whether those beliefs are accurate or not. But the visual system is just one source of input in determining what we come to believe. Thus, even though the visual system works in ways which guarantee that we will be subject to illusions in nonstandard conditions, this does not condemn us to false beliefs.

What is true of the visual system is true of other systems governing perceptual input. They work so well in providing us with accurate impressions of the world around us because the information-processing tasks they perform are made possible by a series of presuppositions about normal environments. When we find ourselves in nonstandard environments, ones which violate the presuppositions of our perceptual systems, we are not fated to mistaken beliefs. We can learn to overcome the misleading impressions which our perceptual systems provide in atypical environments, even if what we learn does not alter the fact that our perceptual systems will provide us with such misleading impressions in those circumstances.

We now have a very general answer to part of our question about how knowledge is possible. How is perceptual knowledge possible? It is possible because our perceptual systems build in certain presuppositions about typical environments, presuppositions which are indeed true of those environments. Such presuppositions make the task of extracting information about the world around us tractable.

Without such presuppositions, the stream of stimulation which our senses provide would not allow us to discriminate between infinitely many different fanciful possibilities about what the world around us might be like. Our sensory experience does not allow us to rule out various nightmare scenarios, but our perceptual systems don't need to rule them out in order to provide us with an accurate picture of the world. The presuppositions built in to our sensory systems are true of normal environments, allowing them to reliably produce true beliefs when we are in such environments, and we are in such environments most of the time. Our perceptual systems are thus instrumental in providing us with a rich body of knowledge about the world.

This leaves us open to misleading appearances when we are not in standard environments, but that does not make false belief inevitable. As we learn more about the world, and about our susceptibility to error, we can learn to compensate for the misleading appearances which our perceptual systems produce.

Some perceptual errors are, indeed, inevitable. Even in environments where our perceptual systems are reliable, they are not perfectly reliable. Nevertheless, the ways in which the perceptual systems are structured to take advantage of deep regularities in our environment ensures that they will typically produce true beliefs.

There are, of course, a great many details which an account of perceptual knowledge needs to fill in. I have sketched how it is that the visual system is able to identify the shapes of various objects around us, but there is a big gap between recognizing that an object is a rectangular solid and recognizing that it is a book. A great deal needs to be said about how we bring our concepts to bear on the objects we see around us. I haven't said anything about that at all. In addition, the ways in which our background beliefs are brought to bear on the processing of perceptual information remains to be explained. A child who is unfamiliar with a certain visual illusion may be wholly taken in by it, forming the mistaken belief that the world is just as it appears. With more experience, the child will learn to recognize illusions for what they are and this new background knowledge will do its work to avoid error without any need for conscious attention or effort. It would be nice to know how our cognitive system accomplishes this feat, and I have said nothing at all about this. What has been offered here, then, is nothing more than the tip of the iceberg, a glimpse of one small but important aspect of how perceptual processing works.

What this account reveals, however, partial as it is, goes a very long way to explaining how knowledge is possible. The input to our perceptual systems can seem remarkably thin. In the case of vision, the image projected on the retina by external objects is only two-dimensional, and that is all the visual system has to go on in figuring out what the world outside it is like. From a certain perspective, this can make knowledge of that world seem impossible to achieve. If we think that the visual system needs to figure out which of all the different imaginable worlds we inhabit, and that determination has to be made on the basis of nothing more than the two-dimensional images the retina provides, perceptual knowledge of the world around us appears impossible. The input to the visual system offers no basis for discriminating between a world of three-dimensional objects persisting through time which maintain stable boundaries; and objects which are constantly changing their shapes as they move; or objects which pop in and out of existence; or any of the other nightmare scenarios we can dream up. And we now see that Descartes's skeptical worries, while they are not presented as challenges to how our visual system might process information, are nevertheless variations on the same theme. Descartes too shows us that there are alternative possibilities for how the world might be, various imaginable worlds we might inhabit, which we cannot discriminate among on the basis of the input which our senses provide. And if we think that the visual system has to perform this task on the basis of such extraordinarily impoverished input, or that we, in thinking about the world, have to self-consciously come up with adequate reasons for ruling out the various nightmare scenarios, then we will quickly be led to skeptical conclusions.

What our picture of a tiny portion of the workings of perceptual processing illustrates, however, is that this is the wrong way to think about perception. The visual system doesn't need to figure out, on the basis of the pattern of illumination on the retina, or on the basis of sensory experience, that we live in a three-dimensional world of a very particular sort. Because the working of the visual system presupposes that, we don't need to find reasons to eliminate bizarre possibilities and nightmare scenarios. And once the task of the visual system is narrowed in this way, a task which seemed impossible now becomes tractable. We are fortunate enough to be endowed with perceptual systems which are well adapted to deep regularities of the world in which they operate. And that is how perceptual knowledge is possible.

3.3. Inference

Not all knowledge, of course, is perceptual knowledge. Our knowledge of the existence of subatomic particles doesn't come from seeing them. We don't know the truths of arithmetic by way of perception. Perception provides us with a rich body of knowledge about the world around us, but we are able to go far beyond the direct deliverances of perception as well. Even knowledge which is somehow grounded in perceptual knowledge, as our knowledge of subatomic particles surely is, requires something more than the simple pickup of information which our perceptual systems provide. An important vehicle for generating additional knowledge is our capacity to draw out the consequences of things we already know by way of inference. But how does inference work, and how does it allow us to expand our bodies of knowledge?

Let us start with a very simple example. Suppose I know that

(1) All women are mortal.

And I also know that

(2) Hypatia is a woman.

Let us forget, for present purposes, how I came to know these things. What I want to focus on here is the fact that, from these two pieces of knowledge, I may safely infer

(3) Hypatia is mortal.

The relationship between (1), (2), and (3) is very important to logicians. The argument which has (1) and (2) as premises and (3) as conclusion is said to be *valid*, and what logicians mean by saying that an argument is valid is this: if the premises of this argument were true, the conclusion would have to be true as well. The premises of a valid argument needn't themselves be true. The validity of an argument does not in any way depend on whether the premises are in fact true. Rather, whether an argument is valid depends on a feature of the relationship between premises and conclusion.

Thus, consider the relationship between (4) and (5), on the one hand, and (6) on the other. That is, consider the argument which has (4) and (5) as premises, and (6) as conclusion.

(4) All birds are reptiles.
(5) Polly is a bird.
(6) Polly is a reptile.

This argument too is valid. It is such that if its premises were true, its conclusion would have to be as well. Obviously, premise (4) isn't true; birds are not reptiles. But the validity of an argument does not depend, as we have said, on whether the premises are in fact true. It depends only on the relationship between its premises and its conclusion. Notice that this argument has something important in common with the previous argument. They have a common form. In particular, they are both instances of this form of argument, where (7) and (8) are premises, and (9) is the conclusion.

(7) All A's are B's.
(8) C is an A.
(9) C is a B.

Illustrating the common form of these two arguments also exhibits what it is about these arguments that makes them valid, and it illustrates this by making the relationship between premises and conclusion explicit. Any argument which has this form will be a valid argument since any premises which have this form will guarantee that, if they were true, the corresponding conclusion would have to be true as well. This makes the inference from premises in such a form to the corresponding conclusion as secure as one could ask for. Whatever the world might be like, however weird it might be, however different from our world, if it is a world in which the premises of such an argument are true, it is also a world in which the conclusion is true. Any world in which all women are mortal and Hypatia is a woman will be one in which Hypatia is mortal. Any world in which all birds are reptiles—obviously we're not talking about our world here!—and Polly is a bird will be one in which Polly is a reptile. The validity of an argument is not due to features peculiar to our world, or to worlds similar to our own. A valid argument is valid not only in our world, but in any world whatsoever. An invalid argument is invalid not only in our world, but in any world whatsoever.

What this means is that if one knows certain statements to be true, and one makes an inference from them by way of a valid argument, the conclusion one thereby draws is guaranteed to be true. The logical guarantee which validity provides is altogether different from the sort of reliability which perception provides, even when it is working just as it should in normal environments. As we have seen, perceptual processes, though reliable, are not perfectly reliable, and in this respect, they contrast with valid inference: valid inference is perfectly reliable; valid inference from true premises guarantees the truth of the conclusion drawn from them.

There is yet another important difference between valid inference and perception. What makes perceptual processes work so well, what makes them so very reliable in normal environments even if less than perfectly reliable, is the way in which those processes exploit certain deep regularities in normal environments. They work exceptionally well in the world we all inhabit, but the way in which they achieve their efficacy in gaining true beliefs would make them ill-suited to operation in worlds very different from our own. Our perceptual processes would not be at all reliable in the shape-shifting world, or the world in which objects pop in and out of existence. This does not make such processes terribly fragile. It's not as if any slight change in the way the world might be would undermine their reliability. Nevertheless, it must be acknowledged that the reliability that perceptual processes achieve is dependent on their operation in worlds which are, in certain important respects, like our own. Valid inference, however, has no such limitation. Drawing out the consequences of true statements will produce true conclusions in absolutely any world at all. It will produce true conclusions in the shape-shifting world; it will produce true conclusions in the world with objects which pop in and out of existence; it will even produce true conclusions in a world in which one is hallucinating, or being deceived by a demon, or at the mercy of a mad neurophysiologist.

Valid inferences allow us to expand our knowledge by drawing out the consequences of other things we know. But we expand our knowledge by way of inferences which are not valid as well. To take a well-worn example, if I observe the sun rise morning after morning after morning and conclude, on this basis, that the sun will rise tomorrow, I have drawn a reasonable conclusion and, if I had not previously believed that the sun will rise tomorrow, I have thereby added to my knowledge. We do, it seems, draw inferences like this. But the fact that the sun has risen on previous mornings does not logically guarantee that the sun will rise tomorrow in the way that a valid argument with true premises guarantees the truth of its conclusion. We can imagine a world in which the sun rises on numerous consecutive occasions and then, for whatever reasons, sunrises altogether cease. The fact that the inference about sunrises does not come with a logical guarantee of truth does not make it an unreasonable inference to draw, nor does it mean that the conclusion drawn cannot amount to knowledge. After all, we gain a good deal of knowledge by way of perception, and, as we've seen, our perceptual beliefs come with no logical guarantee of truth either. In the previous section of this chapter, we saw how it is that perceptual knowledge is possible. What I would like to explain in this section is how inferential knowledge is possible.

If all inference were valid inference, the explanation of the possibility of inferential knowledge would be straightforward. But as we have seen above, we can expand our knowledge by way of inference even where the inferences we make are not valid. And this then raises a problem: how is it possible to gain knowledge by way of inferences which are invalid?

Clearly, not just any invalid inference can serve as a source of knowledge. If I see a man walking toward me and immediately conclude, on the basis of that fact alone, that he is out to harm me, then this inferentially formed belief is obviously not a case of knowledge. Similarly, if I see a woman walking toward me and immediately conclude, on the basis of that fact alone, that I will soon inherit a large sum of money, then this inferentially formed belief too is obviously not a case of knowledge. So no one thinks, and no one should think, that any invalid inference whatsoever may produce knowledge. What stands in need of explanation, however, is how invalid inference could ever be a source of knowledge. Why is it ever legitimate to draw conclusions on the basis of invalid inferences?

I will approach this issue in much the same way that I approached perceptual knowledge. Just as we examined how perception actually works, I want to look at the kinds of inferences we actually make. Much of human inference is just as automatic as perceptual belief is. Our mental machinery goes to work without any need for conscious attention, producing new beliefs on the basis of other things that we believe. Because such inference takes place in us automatically, outside the focus of consciousness, we are in no position to report on these inferences; they are not introspectively available to us. This does not mean, however, that we are in no position to know how such inferences are made. They are the subject of a good deal of experimental work in psychology. By way of controlled experimentation, we can figure out just what kinds of inferences people tend to make. As it turns out, there is a great deal of uniformity in these inferences.

It was not at all obvious, pretheoretically, that there should be such uniformity across different human subjects. One might have thought that different people, with different training, experience, and education, would reason in very different ways. Of course, the beliefs people have are quite different from one another. You may know a great deal about history, but little about physics. Someone else may know a great deal about physics, but very little about history. Given this difference in the background information which various individuals bring to bear on the inferences they make, it is inevitable that when two people are given the same bit of information, they may draw quite different conclusions. But this may simply reflect that difference in background information between the subjects—the premises, as it were, of the inferences they draw—rather than the kinds of inferential connections they are responsive to. If one subject believes, for example, that all birds are reptiles, while a second subject believes that all birds are mammals—our two subjects being astonishingly misinformed about birds!—then it should be no surprise that they will draw different conclusions on being told that Polly is a bird. But they may both be making an inference in accord with the argument form labeled (7), (8), (9) above. In this case, the form of the inference the two subjects make is exactly the same even though they came to completely different conclusions on being given the very same information. Their different conclusions were not due to the structure of the inferences they drew but rather to the different background beliefs they had when making those inferences.

Thus, when subjects given the same bit of information draw different conclusions, it may be that the difference between them is due to the different background beliefs on which they draw in coming to their conclusions, or it may be, instead, that the very structure of the inferences they are making varies from one individual to another. Casual observation will not allow us to determine which of these is actually the case, and it is for this reason that it is not immediately obvious just how much variation there actually is in human inference itself. But although casual observation is not sufficient to capture the structure of the inferences people make, carefully controlled experiments allow us to figure this out. So let us turn to some of the experimental work which has been done in this area.

Beginning in the early 1970's, Amos Tversky and Daniel Kahneman conducted a series of experiments designed to examine the kinds of inferences which people regularly make. Their results presented a picture of human inference which was not flattering. Tversky and Kahneman argued that our "intuitive expectations are governed by a consistent misperception of the world" (1971, 31). Richard Nisbett and Eugene Borgida, commenting on the implications of the literature which began to emerge from this experimental program, remarked that their work had "bleak implications for human rationality" (1975, 935). There seemed to be a remarkable uniformity in the kinds of inferences people make, and those inferences were not, by any reasonable standards, good ones. People reason badly; the inferences we make regularly lead us to false beliefs; we're deeply irrational by our very nature. Inferences of this sort, quite clearly, are not a source of knowledge.

Consider a kind of inference so widely known that long before the work of Tversky and Kahneman, it had a familiar name: the gambler's fallacy. Imagine a fair roulette wheel, divided equally between red and black slots. The wheel is turned, a marble is sent circling around the edge of the wheel, and, as the wheel slows to a halt, the marble eventually settles into one of the slots. You may place a bet that the marble will land on red or that it will land on black. The likelihood that it will land on red, given that the wheel is fair, is 50%; the likelihood that it will land on black, similarly, is 50%. But suppose now that the marble has landed on black three times in a row. The wheel is sent spinning again and as the marble circles, you decide to place a bet. What should you expect? Which is more likely, that it will land on black yet again, or that it will land on red?

The correct answer is that neither is more likely. We have stipulated that the wheel is fair, and what that means is that, on any given turn of the wheel, red and black are equally likely: the likelihood that black will come up on this occasion is thus 50%, and the likelihood that red will come up is also 50%. The fact that black came up three times in a row is just irrelevant. The wheel has no memory of its past performance. It doesn't work to even things out. The current spin of the wheel is like every other: a random spin of a fair wheel with just as many red slots as black.

Despite this fact, a surprisingly large percentage of people think that, in these circumstances, red is more likely to come up. "It's due," as people often say. And this is the gambler's fallacy. Tversky and Kahneman found that people make inferences just like this in a very wide range of circumstances.

Thus, let us suppose that male and female births are exactly equally likely. (They are not, in fact, *exactly* equally likely, but they're close. But we'll assume that they're equally likely for this problem.) Now consider

the last six births at some nearby hospital. Which of these two sequences
do you think is more likely?

(1) G B G B G B

(2) B G B B B B

Kahneman and Tversky 1972 found that more than 82% of subjects
tested thought the first sequence is more likely. But this is just the same
inference as the gambler's fallacy all over again. Consider the first sequence. What is the chance that the first birth at the hospital is, as that
sequence has it, a girl? We've said that boys and girls are equally likely,
so the chance is 50%, or ½. And what are the chances that the second
birth at the hospital, as sequence (1) has it, is a boy? Again, we've stipulated that male and female births are equally likely, so the chance is
50%, or ½. What then are the chances that the first two births, in order,
are a girl and then a boy, as (1) has it? It's 50% of 50%, namely, 25%.
Equivalently, it is ½ of ½, or ¼. And we can continue through the six
births of the sequence. The result is that the likelihood of exactly this
sequence of births is 50% × 50% × 50% × 50% × 50% × 50%, or roughly
1.2%. Equivalently, the result is ½ × ½ × ½ × ½ × ½ × ½, or 1/64. But we
can now see that if we apply the same reasoning to the second sequence,
as we must, we will get exactly the same result. The likelihood that the
next six births in order at the hospital are as sequence (2) indicates is
1/64, or roughly 1.2%. Indeed, *any* sequence of six births, including
G G G G G G is exactly the same. But this is not what subjects judge.

Why do people make this error? What Tversky and Kahneman
found, in conducting numerous experiments of this sort, is that subjects
judge probability by way of what they called *representativeness*. We've
said that male and female births are equally likely. So the process by
which such births are produced is, in relevant respects, utterly random.
Sequence (1) looks representative of the kind of result one would get
from a random process. Sequence (2) looks far less representative. And
a sequence of six female births in a row, or six male births in a row, does
not look remotely representative of the kind of result one would expect
from a random process. Subjects overwhelmingly judge likelihood by
representativeness even though representativeness is not a good indi-
cator of likelihood at all, at least in cases like this one.

Let us look at another set of experiments. Suppose that I want to figure out which of two candidates is likely to win a certain election, and I decide to take a poll. If my poll has a very small sample size—perhaps I just randomly ask a dozen likely voters—then there's very little reason to think that the preferences of those dozen people are likely to reflect the preferences of the far larger electorate. Notice, for example, that if I find that seven out of the twelve people I ask prefer candidate A to candidate B, my result, reported in percentage terms, would indicate that candidate A is preferred by roughly 58% of my sample, and candidate B is preferred by roughly 42% of those asked. This is a sixteen-percentage point margin, which, in electoral politics, is huge. Any politician running for office would sleep well at night knowing that they have a sixteen-point lead. But a change of just one individual vote in my sample could change those percentages to a 50/50 split—not exactly the sort of result which leaves politicians sleeping well. Now no one, of course, would predict electoral results on the basis of such a small sample of opinions. If one is to have any confidence in the result of one's poll as a basis for predicting the election, one needs a far larger sample of voters. This simple point is what probability theorists refer to as the Law of Large Numbers: the larger one's sample size, the more likely it is to correctly reflect features of the population it samples.

Tversky and Kahneman were interested in seeing the extent to which our everyday expectations and estimates of likelihood take the Law of Large Numbers to heart. Their results were, to say the least, disappointing: "People's intuitions about random sampling appear to satisfy the law of small numbers, which asserts that the law of large numbers applies to small numbers as well" (1971, 25). In case after case, through a long series of ingenious experiments, Tversky and Kahneman documented a wide range of ways in which human inference goes wrong, adding more and more evidence for their indictment of human reasoning. We "commit a multitude of sins against the logic of statistical inference" (1971, 31). We are, it seems, on their view, not just very bad reasoners; we are built to be bad reasoners. It is worth quoting, once again, their overall assessment: our "intuitive expectations are governed by a consistent misperception of the world" (1971, 31).

Notice that Descartes's skeptical worries pale by comparison with this conclusion. Descartes's nightmare scenarios raised serious questions about whether knowledge is possible, but he did not argue that our beliefs are in fact largely false. The point of the troubling scenarios which Descartes raised was that *for all we know* our beliefs *might* be false; we are in no position to tell whether our beliefs accurately represent the world as it is or, instead, get things horribly wrong. That is, to be sure, a pessimistic view about the human capacity for knowledge. But Tversky and Kahneman make Descartes look like an optimist, for they argue that our expectations provide us with "a consistent misperception of the world." And this would mean, if they were right, not that, for all we know, our beliefs *might* be false. It would mean that, for the very most part, our beliefs *are* in fact false.

There is something deeply suspicious about any such conclusion. Are Tversky and Kahneman really arguing that the view we have of the world around us is almost entirely mistaken? Consider our survey of the phenomenon of knowledge in Chapter 2. Consider, in particular, the wide range of beliefs we form as a product of inference. We all believe the sun will rise tomorrow. We believe that when we leave our houses and apartments for whatever reason, we will be able to find our way back home. We believe that the cities and towns we live in will persist, and that tomorrow they will look very much the same as they do today. We believe that cars will continue to move through the streets, and that birds will continue to fly through the air. We believe that the food we eat will, for the very most part, continue to sustain us, and that water will continue to quench our thirst. We have a rich set of expectations about our friends, and loved ones, and acquaintances; about their appearance, and their behavior, and their characters. Are Tversky and Kahneman claiming that the great majority of these expectations are mistaken? That is, quite literally, beyond belief.

But that is not all. In addition to these common-sense beliefs we all have about the world, beliefs which are a product of our intuitive expectations, there are the beliefs we have as a product of modern science. The sciences have given us a very rich picture of the world around us, a picture which is highly confirmed by rigorous experimentation as well as by the success of various technological applications of scientific theories. Those of us who are not scientists, and who have not performed these experiments, or even know the experimental evidence for the current scientific worldview, nevertheless know a good deal about the conclusions which our best current theories have reached, and we accept a good part of that worldview as true. Are we supposed to believe, if we accept Tversky and Kahneman's conclusions, that the scientific picture of the world is largely mistaken? Any such conclusion is, to be sure, frankly unbelievable.

Just as the skeptical conclusion which emerged from Descartes's parade of nightmare scenarios could not be accepted even before we could say just where his argumentation went wrong, we should approach the very sweeping claims which Tversky and Kahneman made in many of these early papers with a very large grain of salt. We should suspect that there is something wrong, indeed, fundamentally wrong, with any argument which leads to such conclusions, even before we can say where the error lies. This does not relieve us of the responsibility of figuring out where they go wrong, and I mean to address that issue directly. It is, nevertheless, important to approach that task with a firm grasp on reality.

The key to seeing the problem with Tversky and Kahneman's approach lies in their remark that their work shows that we "commit a multitude of sins against the logic of statistical inference." Our inferential tendencies are being assessed by comparison with the dictates of probability theory, and we are found to come up short in case after case after case. We do not reason in the ways that probability theory seems to suggest we should.

Now in certain respects, this is not terribly surprising. Most people have no training at all in probability theory, so it is no great surprise that we don't make inferences which conform to a theory of which we are ignorant. But the fact that this is unsurprising, at least at this level of generality, does not yet show any problem with Tversky and Kahneman's arguments, and for two different reasons.

First, their results would be far less surprising if what they found was that those who are unacquainted with probability and statistics do not reason in ways which conform to the logic of statistical inference, but those who are acquainted with these formal theories do. But this is not what Tversky and Kahneman found. When they compared naive subjects—that is, subjects with no training in probability theory—with advanced students in these areas, as well as with professionals who regularly make use of probability theory in their work, they found surprisingly little difference in the inferential behavior of the different groups. "Evidently," they concluded, "statistical training alone does not change fundamental intuitions about uncertainty" (1973, 68). Advanced students and professionals in these areas performed only very slightly better on many of these tasks than naive subjects. What Tversky and Kahneman found, therefore, was not remotely like pointing out that, for example, if you ask random people about quantum mechanics, most of them will prove to be entirely ignorant about it. In that case, those with special training would answer questions about quantum mechanics with ease. So part of what is surprising about the Tversky and Kahneman results is that even those who have the relevant training typically fail to bring it to bear on much of their cognitive lives. That is a very interesting, surprising, and disappointing result.

But second, even if the performance of statisticians and probability theorists on these problems were far, far better than, in fact, it is, and even if it showed that they regularly reason in ways that those ignorant in these areas fail to do, that would hardly let naive subjects off the hook. If most of us reason very badly because we lack training in these areas, and that leaves us with a completely inaccurate picture of the world, then our ignorance and errors are not irremediable. We could all go out and take courses in the relevant areas, and thereby join the ranks of the enlightened. We would still, however, have a completely inaccurate picture of the world until we take those courses, a picture which is a product of our ignorance and errors. So pointing out that it's no surprise that most of us, untrained as we are in probability theory and statistics, do not conform to its dictates just doesn't help if our reasoning is problematic to the extent that it fails to live up to the standards of those fields.

The real issue here, then, is whether failing to satisfy the standards defined by probability theory constitutes an epistemic sin, as Tversky and Kahneman claim, and, in particular, whether by failing to meet those standards we inevitably end up with a radically mistaken view of the world around us.[2] What I want to argue is that Tversky and Kahneman impose an unreasonable standard on human inference.

The logic of statistical inference does not set the standard for good reasoning, as I will argue. Further, just to head off an obvious worry: I will not be arguing that they set the bar for good reasoning too high, and that we should lower our standards; I will not be arguing that even if we get a C– on the tests that Tversky and Kahneman have given us, we should not feel bad because, after all, a C– isn't such a bad grade. Rather, I will argue that the standards they measure human inference against are just the wrong standards. These standards do not define good reasoning. What makes an inference a good inference is not, I will argue, what Tversky and Kahneman take it to be.

Tversky and Kahneman do not ever argue that the standards of statistical inference set the proper standards for good reasoning; they simply assume it. In addition, they also assume that if one regularly violates the standards of statistical reasoning, one will end up with far more false beliefs than true ones—"a consistent misperception of the world." Both of these claims are, I believe, mistaken.

The standards of statistical inference, such as the Law of Large Numbers, are not responsive to features peculiar to the world we inhabit. The mechanisms which govern perceptual appearances, on the other hand, as we saw, make certain presuppositions about the world in which they operate. Those presuppositions are not true, or even approximately true, of every imaginable world. They are, however, at least approximately true of standard environments in our world, and it is this fit between the presuppositions of our perceptual mechanisms and features of standard earthly environments which explains how those mechanisms contribute so reliably to the production of true belief. If we were to reason by way of the standards of statistical inference, however, our successes would not be explained in the same way as the successes of our perceptual mechanisms. The Law of Large Numbers is meant to define reasonable inference in any world whatsoever. For that reason, statisticians and probability theorists do not defend their appeal to the Law of Large Numbers by trying to show that there is some feature of our environment which it accurately presupposes. Conforming to the Law of Large Numbers would be a good way to reason, it is thought, whatever the world might be like.

Consider, however, the following possibility. Imagine that our inferential mechanisms work in ways which are structurally similar to the ways in which our perceptual mechanisms work. Imagine, that is, that we have native inferential tendencies to draw conclusions in ways which make certain presuppositions about standard environments. Such inferential tendencies would not reliably produce true beliefs regardless of what the world might be like. Instead, they would reliably produce true beliefs in the world we inhabit because the presuppositions built in to the inferential mechanisms are ones which are at least approximately true of our world. Our inferences would work well because of the fit between those presuppositions and features of the world in which they operate. If our inferential mechanisms worked in this way, they would reliably produce true beliefs in our world, even though they would not work at all well in worlds sufficiently different from ours. That wouldn't be such a bad thing, any more than the fact that our perceptual mechanisms would not work well in a world populated by objects which pop in and out of existence shows that we are badly served by having the perceptual mechanisms that we do.

I will argue that much of our inferential behavior is of exactly this sort. It is well adapted to the environments in which it typically operates and so it tends to produce true beliefs. Much of the way in which we reason would not work well in worlds very different from our own, but it is none the worse for that. A good reasoning mechanism is one which reliably delivers the goods: it is one which tends to produce true beliefs. Just as a good screwdriver needn't work well in every imaginable world, but only in ours, a good inferential mechanism needn't work well in every imaginable world either.

This is not the way in which most people think about good inference. If one's model of good inference comes from logic, and one's paradigm of good reasoning is drawing the conclusion that Hypatia is mortal from the premises that all women are mortal and Hypatia is a woman, then the way of thinking about inference that I am proposing here will seem quite unnatural. Similarly, if one has training in statistical inference and one has been taught to think about the Law of Large Numbers as paradigmatic of good reasoning, the approach I propose here will be quite foreign to the way one is accustomed to thinking about how inference ought to proceed. The suggestion I am making, however, is that our inferential mechanisms embody a solution to the difficult problem of drawing accurate conclusions about the world in which they operate by making substantive presuppositions about that world. Our inferential mechanisms, if I am right, are tailored to certain pervasive features of our world, and they work well because they are adapted to our world in just this way.

I cannot possibly make the case for such a large claim in any comprehensive way. Rather, what I propose to do here is offer a proof of concept. I will take a central case of the kind of inference Tversky and Kahneman object to and argue that, far from illustrating a defect in our reasoning, it illustrates a way in which the inferences we make work well because of the structure they share with our perceptual mechanisms: they are tailored to operate in the environments in which we are usually found.

If I can show that there are inferences we make which violate the standards Tversky and Kahneman take for granted, and yet are none the worse for that, then this will go some substantial distance toward a defense of my way of looking at a theory of good inference. Because they lay such stress on it, and because their approach in this particular case is so very appealing, I will take a closer look at our tendency to violate the Law of Large Numbers, what Tversky and Kahneman refer to, mockingly, as the psychological law of small numbers. These inferences are not only far better than Tversky and Kahneman allow; they are, I will argue, highly reliable sources of true belief, and, for that very reason, a paradigm of good, rather than bad, reasoning.

As Tversky and Kahneman emphasize, we routinely violate the Law of Large Numbers, making predictions on the basis of very small samples. This tendency is especially apparent in the case of vividly presented information. Case studies with a great deal of detail, rich in the imagery they provoke, are far more likely to be remembered than statistical summaries of large samples, which readers and listeners often find hard to pay attention to. More than this, a vividly detailed example, whether described, or, even more so, directly encountered in one's personal experience, has a far greater tendency to drive one's inferences. (See, e.g., Nisbett and Ross, 1980, chapter 3.) Nisbett, Borgida, Crandall, and Reed (1976, 129) give a nice example of this contrast between vividly presented information and statistical summaries.

Let us suppose that you wish to buy a new car and have decided that on grounds of economy and longevity you want to purchase one of those stalwart, middle-class Swedish cars—either a Volvo or a Saab. As a prudent and sensible buyer, you go to *Consumer Reports*, which informs you that the consensus of their experts is that the Volvo is mechanically superior, and the consensus of the readership is that the Volvo has the better repair record. Armed with that information, you decide to go and strike a bargain with the Volvo dealer before the week is out. In the interim, however, you go to a cocktail party where you announce this intention to an acquaintance. He reacts with disbelief and alarm: "A Volvo! You've got to be kidding. My brother-in-law had a Volvo. First, that fancy fuel injection computer thing went out. 250 bucks. Next he started having trouble with the rear end. Had to replace it. Then the transmission and the clutch. Finally sold it in three years for junk."

Almost everyone who reads this anecdote not only recognizes this sort of situation as one they themselves have been in, but knows equally as well that they would be influenced by the vivid story far more than the theory of probability tells us we should be. If *Consumer Reports* surveyed one thousand Volvo owners and the vast majority of them reported that their car had been quite reliable, then the vivid anecdote should be given 1/1000 the weight of the information in the consumer magazine. What most people recognize, however, is that they would not be so coldly analytical in this situation, and this is well confirmed experimentally by Tversky and Kahneman as well as by Nisbett et al. The single vivid case swamps the more telling statistical information when it comes to what influences our opinions, and, of course, without the statistical information presenting contrary evidence, the single vivid case is often fully sufficient to affect one's decisions and beliefs. This is Tversky and Kahneman's psychological law of small numbers gone wild: a sample size of one is sufficient to drive the inferences we make about the population from which that sample is drawn.

Just how bad is it to draw inferences about a population on the basis of small samples? In particular, how bad is it to draw inferences about a population on the basis of a single case? This is, of course, the most extreme violation of the Law of Large Numbers one could ask for, and so one might think that someone who reasons in this way is likely to do far worse than someone who responsibly gathers information about a great many cases before reaching any conclusion. If the conclusion is about the next individual within the population one will encounter, however, the difference between the results of the person who jumps to conclusions on the basis of a single case and the person who gathers extensive evidence is far smaller than one might expect.

Imagine that we have an urn filled with ping pong balls. The various balls in the urn may be black or white, but not any other color. We don't know in advance what the ratio of black balls to white ones is; indeed, for all we know, the urn could be filled entirely with white balls or entirely with black balls. Two individuals, the Hasty Generalizer and the Methodical Reasoner, are each asked to guess what color the next ball drawn from the urn will be. The Hasty Generalizer, true to his name, picks a single ball from the urn, and then replaces it. If the ball he picked is black, he predicts that the next ball drawn from the urn will be black. If the ball is white, he predicts that the next ball drawn will be white. The Methodical Reasoner, on the other hand, completely empties out the urn, counting every last ball it contains before replacing them. If the majority of the balls are black, she predicts that the next ball drawn from the urn will be black. If the majority are white, she predicts that the next ball drawn will be white. And if there are exactly as many black balls as white, she flips a coin to determine her prediction: if it comes up heads, she predicts white, and if it comes up tails she predicts black.

Tversky and Kahneman, of course, would favor the strategy of the Methodical Reasoner and heap scorn on the Hasty Generalizer. But how much worse will the Hasty Generalizer do? It depends on the ratio of black balls to white in the urn. In the extreme cases, in which the urn is filled entirely with balls of one color, or the urn contains exactly the same number of black and white balls, the Hasty Generalizer will be exactly as accurate as the Methodical Reasoner. If, for example, the urn is entirely filled with black balls, the Hasty Generalizer will inevitably pick a black ball when he chooses his single sample, and he will then predict that the next ball chosen will be black. And he'll be right, since the urn is filled with nothing but black balls. He does exactly as well as the Methodical Reasoner, who counts every ball in the urn, noticing that they are all black, before making exactly the same predication. On the other hand, if the urn has the same number of black and white balls, the Hasty Reasoner will get the right result only half of the time. He is as likely to pick a black ball as a white one for his sample case, and, after replacing it, is as likely to pick a black ball as a white one on the next try. But the careful enumeration of the Methodical Reasoner does not allow her to do any better. They both are successful only half of the time.

What then of cases in between the extremes? It is here that the differences between these two strategies emerge. Consider then the case in which 60% of the balls are black and 40% are white. The Methodical Reasoner will examine them all, and on finding that the majority are black, she will predict that the next ball drawn from the urn will be black. She'll be right 60% of the time. The Hasty Generalizer will draw a single ball, and 60% of the time the ball he draws will be black; 40% of the time it will be white. There are two different ways in which the Hasty Generalizer may accurately predict the color of the next ball: he will be right either if he draws a black ball as his sample and then, after replacement, draws another black ball, or, alternatively, if draw a white ball as his sample, and then, after replacement, draws another white ball. The chances that he will draw black two times in a row are 60% of 60%, or 36%; the chances that he will draw white two times in a row are 40% of 40%, or 16%. So the chances that the Hasty Generalizer will make the right prediction in this case is the sum of these two percentages, or 52%. The Methodical Reasoner does better, but not by much; the Methodical Reasoner does better in only 8% of the cases. The largest difference emerges when the ratio of black balls to white, or white balls to black, is 3:1. In that case, the Methodical Reasoner does better in 12.5% of the cases. This too is not a very large improvement, and if there are a large number of balls in the urn, the Methodical Reasoner's strategy will take a great deal of time to implement. For all that additional work, one would hope for a more substantial improvement over the quick-and-dirty procedure used by the Hasty Generalizer. (For discussion, see Nisbett and Ross 1980, 256–60.) If a great deal hangs on this prediction, it may be worth the additional work to gain a small bit of accuracy. By and large, however, the strategy of the Hasty Generalizer produces results which are nearly as good as those of the Methodical Reasoner, with far less investment of time and energy. All things considered, the strategy of the Hasty Generalizer, at least in these kinds of cases, is not clearly a bad one; indeed, it is often better.

I want to look, however, at far less idealized cases because it is here, I believe, that we may come to truly understand what is at work in the inferences most of us make in everyday situations. We need to examine cases, not of balls in urns, but the kinds of situations most of us encounter in ordinary life. Even if we assume, as a result of the research by Tversky and Kahneman, that there is a tendency for people to draw inferences on the basis of small samples, or even on the basis of a single case, we need to look at what kinds of inferences people tend to draw from such small samples.

Earlier, we looked at the logical form of valid inferences and saw that we could describe certain forms of argument such that any argument in that form is valid. As we noted, any argument with premises having the form of (7) and (8), and conclusion in the form of (9) is valid.

(7) All A's are B's.
(8) C is an A.
(9) C is a B.

It doesn't matter whether an argument in this form is about the mortality of women or the relationship between being a bird and being a reptile; any argument which has this form is such that, if the premises were true, the conclusion would have to be as well. That is what makes arguments in this form valid, and that is what makes an inference from premises in this form to conclusions in the corresponding form a good inference.

Inferences from a sample to a population, or from a sample to the next encountered instance of that population, don't work this way. Even in cases involving large samples of just the sort Tversky and Kahneman would approve of, the good quality of the inference cannot be captured by looking at its form. Thus, for example, consider the following argument.

(10) In a very large sample of robins, virtually all of them were able to fly.

(11) The next robin I see will be able to fly.

It seems quite reasonable to draw the conclusion (11) from (10), and if one were to characterize the form of this inference, it would look like this.

(12) In a very large sample of A's, virtually all of them were B.

(13) The next A I see will be B.

Remember that when we looked at the nature of valid arguments, we saw that validity is a matter of logical form, and what that means is that in valid argument forms, such as the argument from (7) and (8) to (9), one can substitute anything one likes for A, B, and C and the argument will remain valid. One can make the premises of an instance of that argument form false with certain choices for A, B, and C, but one cannot make any instance of that argument form invalid; any instance of that form will always be such that, *if* the premises were true, the conclusion would have to be as well.

But now consider the following instance of the argument form with premise (12) and conclusion (13).

(14) In a very large sample of moments of my life, virtually all of them have been prior to January 1, 2021.

(15) The next moment of my life I see will be prior to January 1, 2021.

Imagine someone noticing that (14) is true about him, and he notices this as the clock is about to strike midnight on New Year's Eve, ushering in the year 2021. The inference from (14) to (15) would be an absurd inference to draw on such an occasion. This is not a remotely reasonable argument, and yet it has the same form as the argument from (10) to (11).

What should we conclude from this? Although validity is a matter of logical form, the arguments we are now looking at are ones which are not valid. Unlike valid arguments, what makes the argument from (10) to (11) a good one cannot be found in its form.[3] Arguments like this may be reasonable; the truth of the premises may give us good reason to believe the conclusion even if, unlike valid arguments, they do not absolutely guarantee its truth. But even a good argument of this kind, such as the argument from (10) to (11), will be such that there are other arguments of exactly the same form, such as the argument from (14) to (15), which are not good at all.

Tversky and Kahneman were not investigating what it is that makes an argument a good argument. They were investigating what kinds of inferences people have a tendency to make. But the fact that people will be very much inclined to infer (11) from (10), does not, of course, mean that they will be even remotely inclined to conclude that (15) is true on the basis of (14). No sane individual would make that inference.

When Tversky and Kahneman argue that there is a psychological law of small numbers, then, although they provide a good deal of evidence that very small samples, even a sample of a single case, will frequently lead people to draw a conclusion about the next sample of the population they encounter, or even about the population as a whole, this cannot mean that people will make such inferences about any category whatsoever. People are certainly quite likely to conclude that the next robin they see will fly even on the basis of seeing only a single robin fly; they are likely to conclude, as well, that all robins fly on the basis of seeing a single robin fly. No one is likely to conclude, however, that all of their meetings with some new acquaintance will be on a Tuesday on the basis of first meeting that person on a Tuesday. No one is likely to conclude that it is always sunny in Paris on the basis of arriving in Paris for the first time on a sunny day. No one is likely to conclude that they will always encounter a certain friend of theirs every time they go to the new restaurant in town on the basis of running into that friend the first time they go to that restaurant. So even if we do, as Tversky and Kahneman point out, have a tendency to draw conclusions about populations on the basis of tiny samples, we are very selective about when we do this; we don't just do it across the board. And this, of course, raises the question of when it is that we draw such inferences. Without knowing when we're inclined to draw inferences about a population on the basis of a small sample, we can't say whether this inferential tendency of ours is a good one or a bad one.

Certain categories are highly uniform with respect to certain properties. If I discover that a certain sample of copper melts at a given temperature, it's a very good bet that others will do so as well. If I find that one baby koala eats eucalyptus leaves, it's a good bet that other baby koalas will also enjoy those leaves. And if the first oak tree I encounter drops acorns, it's a good bet that other oak trees will drop acorns too. All of these conclusions are ones which we are naturally inclined to draw. But the fact that these categories are uniform in certain respects does not mean, of course, that they are uniform in all respects. If the first sample of copper I encounter was owned by my brother, it would be unwise for me conclude that my brother owns all of the copper there is. If my first encounter with a koala was at the San Diego Zoo, it would be unwise to conclude that all koalas are found in that zoo. And if the first oak tree I ever saw was planted by my mother, it would be unwise to conclude that she had planted all of the oak trees there are. These categories are not uniform with respect to these properties, and, of course, no one would be inclined to draw inferences like these.

To what extent are we appropriately attuned to categories which have the sort of uniformity which lends itself to successfully making inferences about the category on the basis of small samples? And to what extent are we sufficiently sensitive to the kinds of properties within these categories that are, indeed, largely uniform in those categories? If we knew the answer to these questions, we might have some idea about whether our tendency to draw conclusions about a population on the basis of small samples leads predominantly to true beliefs.

As it turns out, there has been a good deal of work on cognitive development in children which helps us provide the beginning of an answer to these questions. Susan Gelman 2003 was interested in how very young children categorize various kinds of objects in the world around them. For example, what is it that makes a child think of two different animals it sees as being of the same kind? Clearly, there is something about the visual appearance of an animal that is relevant here, but we should be able to say something more than that classification depends on how things look. Interestingly, Gelman found that how things look is not decisive in determining a child's classificatory behavior.

Adults, of course, are well aware that appearances may be misleading, and that proper classification does not depend exclusively on how things look. We recognize that not everything that looks like water is water; gin and vodka look just like water, as do a number of poisonous liquids. What makes something water is not a matter of its appearance, but its chemical composition, and even if we are not often, or, for most of us, ever, in a position to test a substance for its chemical composition, we know full well that that appearances present only a rough guide to the nature of different substances, and things which look alike may actually be quite different. All that glitters is not gold. Nor can plants and animals be categorized on the basis of appearance alone. Whales live in the ocean and in many respects resemble some very large fish, but they are not fish; they are mammals. What determines the proper classification of these various kinds is not their superficial appearances, even if that is often sufficient for recognizing them; it is various underlying characteristics, such as their chemical composition, their genetic makeup or evolutionary heritage, and so on, in virtue of which they bear deep similarities that govern their behavior in lawful ways. Philosophers refer to these categories as natural kinds, and a good deal of our classificatory behavior, and our classificatory language and thought, pick out such kinds even before we know exactly what it is that makes a given kind the kind it is. Human beings thought and talked about water long before they knew that what makes something water is its chemical makeup, let alone that it is H_2O. We were able to do this because we had a fairly good recognitional capacity for picking out water from other substances, even as we recognized, if only implicitly, that the features on the basis of which we were able to pick it out do not make it the kind of substance it is. And the same is true of other natural kinds. I'm pretty good at recognizing a bear when I see one, but I know that just as whales look like large fish but aren't really fish, there may well be various animals that look like bears but aren't bears. I don't know what features of an animal make it a bear; that is, I don't know what the underlying properties of the species are which govern the category, but, at least in the regions I tend to travel in, that doesn't get in the way of my being able quite reliably to recognize bears when I see them.

Our ability to recognize natural kinds in our environment even before we know just what makes them the kinds they are also allows us to make inferences about other members of these kinds on the basis of acquaintance with very few instances of the kind. The underlying characteristics of a kind ensure a certain amount of uniformity: one sample of water, or copper, or gold will interact with other substances in much the same way as any other sample of those kinds. And, to a first approximation, the same is true of other natural kind categories. The inferences we make in this way are not perfectly accurate. We do make mistakes. But the way in which the underlying features of the kind lawfully ensure uniformity also ensures that certain inferences, even on the basis of very small samples, will be reliable.

Adults are thus able to make inferences about natural kinds in a reliable way because natural kinds are, by their very nature, uniform in certain properties. One might have thought that this sensitivity which we show to that uniformity is the product of the extensive experience any adult has with the natural world. That is, one might have thought that our responsiveness to the uniformity of natural kinds is something that we learn. What Susan Gelman has shown, building on the work of a good many other developmental psychologists, is that this responsiveness to natural kinds is not, in fact, learned; it is innate.

From their earliest classificatory behavior, children show that their way of conceptualizing natural kind categories presupposes that it is various underlying features of these categories that make them what they are, rather than their superficial appearances. Children recognize that making an animal of one kind look like an animal of another kind doesn't make it that other kind of animal. Painting stripes, for example, on a horse doesn't make it a zebra. And when told that two animals that look very much alike are actually of different kinds—for example, that bats are not birds but mammals—children immediately conclude that the insides of these animals and much of their behavior will best be predicted by way of their kind membership rather than by way of their appearance: the bats will have insides like other mammals rather than like various birds. Gelman argues that children are what she calls "psychological essentialists": from the very beginning, children implicitly presuppose in their way of thinking about the natural world that it comes divided into kinds defined by underlying features of which they may be unaware, and that the superficial appearances of these kinds is only a first-pass approximation to the real natures of these natural kinds. Children cannot, of course, articulate this view. They are not aware of making any such presupposition. Indeed, the conceptual capacity to articulate such a view does not appear in children until long after they have been making this presupposition in their thoughts, and inferences, and actions. It is this presupposition, however, which allows human beings to make such reliable inferences about the natural world.

And the same is true of adults: adults need not be able to articulate this view of natural kinds; they need not explicitly believe it; nevertheless, it is presupposed in the way we classify the denizens of the natural world, the way we conceptualize the various categories of objects in the natural world, and the inferences we make about the members of those categories. What makes these inferences reliable is that the natural world actually has categories of objects which have the kinds of deep law-governed regularities characteristic of natural kinds, and that we are sufficiently responsive to these regularities in our interactions with the world. We are psychologically constituted, from birth, in a way that mirrors the structure of natural kinds.

What this means is that the way in which we make certain inferences, while it is very reliable, would not be reliable in worlds very different from our own. In a world which had no natural kinds, in a world in which there were no deep law-governed regularities, our inferences would be extremely unreliable. Even in a world governed by deep lawful regularities, if the superficial properties of natural categories which attract our attention did not even roughly allow us to reliably sort members of those categories into their natural kinds, the inferences we are inclined by our natures to make about those natural categories would be extremely unreliable. Our inferential tendencies are good ones to have not because they would be good in any imaginable world whatsoever, but because they are well adapted to deep-seated lawful regularities of the world we inhabit. And in this respect, what makes these inferences good ones to draw is explained in the very same way as the reliability of our perceptual judgments rather than the reliability of valid inferences. Valid inferences, such as the inference from the premises that all women are mortal and Hypatia is a woman to the conclusion that Hypatia is mortal, are reliable, indeed, perfectly reliable, in every imaginable world, even a world in which we are constantly hallucinating or being deceived by an evil demon or a mad neuroscientist. Many of the inferences we make about natural kind categories, however, are good ones to make in virtue of being well adapted to our world, even if they would not be good inferences in every imaginable world. It is by the grace of this fit between the presuppositions of our inferential behavior and the natural kind structure of our world that we are able to achieve so much of the knowledge we have.[4]

I am not saying that we never make valid inferences—inferences which are such that, if the premises were true, the conclusions would have to be true as well. Inferences of that sort work well no matter what the world might be like. What I have been arguing, however, is that much of our knowledge is achieved by way of inferences which are not like that at all, but instead are good inferences because their very workings presuppose that the world in which they operate has a certain structure—the structure characteristic of natural kinds—and, more than this, the world does in fact have that structure.

3.4. Conclusion

In this chapter, we have examined two important sources of knowledge: perception and inference. We have seen that the way in which our perceptual systems work would not allow them to reliably produce true beliefs in just any imaginable environment. Rather, and perhaps unsurprisingly, our perceptual systems are adapted to certain stable and widespread features of the environment in which we live. We come to understand how our perceptual systems work by seeing how they can be made to go wrong; this is the reason why psychologists who study perception are so interested in visual illusions. When we see how the visual system can be fooled, we thereby come to understand what the visual system presupposes in its ordinary interactions with the environment. It is by making these presuppositions, presuppositions which are true of typical environments in which we tend to be found, that we are able so quickly and easily to pick up information about the world around us. The visual system doesn't need to learn these things about typical environments. Rather, these presuppositions are simply built in to the visual system from the very start. It is in virtue of having a visual system which is attuned to stable features of the environment in this way that we are able to gain perceptual knowledge.

The reliability of the inferences we make, it might seem, is to be explained in a different manner. If one's model of inference is valid argumentation, arguments in which the conclusion must be true if the premises were true, then the quality of our inferences would be ensured whatever environment they might operate in. Such inferences, therefore, need not be attuned, as our perceptual systems are, to features peculiar to our environment: a good inference in one world would be a good inference in all. What I have argued, however, is that a good deal of our inferential knowledge is, surprisingly, to be explained in exactly the same manner as our perceptual knowledge. Much of our inferential behavior is, like perception, attuned to stable features of typical environments in which it occurs. What this means is that our inferential successes, like the successes of our perceptual systems, are the product of an adaptive fit between the presuppositions built in to the inferences we are naturally disposed to make and the typical environments in which we make them. We do not need to learn these things about our environment; our natural inferential dispositions simply build in these presuppositions from the very beginning.

Much as perception and inference are central parts of our cognitive machinery, they do not, of course, constitute the whole of it. Nevertheless, this quick tour of these two important parts of our knowledge-gathering mental equipment provides us with important insights in understanding how knowledge is possible.

Suggestions for Further Reading

A good introduction to work on visual perception can be found in Nicholas Wade and Mike Swanston's *Visual Perception: An Introduction*, Psychology Press, 2013. A far more advanced, but extremely important, approach may be found in David Marr, *Vision*, W. H. Freeman, 1982. Tversky and Kahneman's important early contributions to the study of inference, along with a number of papers by others, may be found in Daniel Kahneman, Paul Slovic, and Amos Tversky, *Judgment under Uncertainty: Heuristics and Biases*, Cambridge University Press, 1982. Kahneman's more recent thinking on these matters, which is in many important ways quite different from that found in his early papers, is well presented for a general audience in his *Thinking, Fast and Slow*, Farrar, Strauss and Giroux, 2011. A popular account of the collaboration between Tversky and Kahneman may be found in Michael Lewis, *The Undoing Project: A Friendship That Changed Our Minds*, W. W. Norton, 2017.

Work on conceptual development in children and its connection with natural kinds may be found in Frank Keil, *Concepts, Kinds, and Conceptual Development*, MIT Press, 1989; Susan Gelman, *The Essential Child: Origins of Essentialism in Everyday Thought*, Oxford University Press, 2003; and Ellen Markman, *Categorization and Naming in Children: Problems of Induction*, MIT Press, 1989.

Important philosophical work on natural kinds may be found in W. V. Quine's seminal paper "Natural Kinds," in his *Ontological Relativity and Other Essays*, Columbia University Press, 1969.

I have developed the view presented in section 3.3 in greater detail in *Inductive Inference and Its Natural Ground*, MIT Press, 1993.

4

Knowledge from the Inside

The First-Person Perspective

4.1. What the First-Person Perspective Has to Offer

In the last chapter, we looked at two important processes which can, at times, operate within us automatically, without the need for any self-conscious attention or supervision on our part. These automatic processes go to work within us, and they are often, indeed typically, a source of knowledge. Because these processes operate outside the scope of conscious attention, they can only be examined through a third-person perspective, and that is exactly what experimental psychology offers us.

Our intellectual lives, however, are often vividly present to mind. We self-consciously think about what we ought to believe, or whether we ought to go on believing as we do. At these times, we have a first-person perspective on belief acquisition and revision. These processes do not simply take place within us, in the way that the processes of digestion and the circulation of the blood and so many other bodily processes do. When we stop to reflect on what to believe, we are not just directly aware of mental processes going on within us. We are not mere observers of our mental processes. We are, instead, actively involved in making those processes take place. We are, in a word, epistemic agents. Or so it seems.

The first-person perspective on our intellectual lives is thus important, in the first place, because adult human beings, unlike young children and nonhuman animals, may take charge of their epistemic lives. Our beliefs do not just wash over us, or, at least, not always. Our capacity to reflect on our beliefs and on what we ought to believe allows us to play an active role in belief acquisition and revision. That active role, when we are

Scientific Epistemology. Hilary Kornblith, Oxford University Press. © Oxford University Press 2021.
DOI: 10.1093/oso/9780197609552.003.0004

playing it, seems to be fully present to consciousness; indeed, it is because we are, at these times, actively engaged in managing our epistemic lives that we cannot fail to be aware of the role we play.[1] This ensures that the first-person perspective, the perspective that adult human believers have on their own mental lives by way of introspection, offers a perspective on our lives as believers and knowers that gives us unique insight into the very nature of our ability to achieve a kind of knowledge of which less so-phisticated knowers are incapable. It is little wonder that epistemologists have, for so long, viewed the first-person perspective as a crucial source of epistemological understanding.[2]

This is not, however, the only reason why the first-person perspective on belief acquisition and revision has played such a central role in episte-mological theorizing. When we wonder about what we ought to believe, or whether we are believing as we should, we find ourselves in need of epistemic guidance.[3] We want to know not only what we should believe, but what we should do in order to believe as we should. Should we consult with others to see what they believe on the matter in question? Should we, instead, just try to think things through on our own? Should we re-ex-amine the evidence we have to see what it really supports, or should we be occupying ourselves with further evidence gathering before we even try to make some assessment of where the evidence leads? Even if we think, in the end, that consultation with others is what is called for in trying to determine what we should believe, that decision to seek consultation is our decision, not someone else's. It is up to us whether we consult with others, and that is a decision which cannot be left to others. When we reflect on what to believe, we are in charge; we make the decisions. The guidance we seek, therefore, must, in the end, be guidance we settle on for ourselves. Ultimately, we decide what will guide our intellectual lives. This guidance, these decisions about how to proceed when we wonder how to proceed, must come from within. Our decisions on these matters take place under conditions of reflection; they are made from the first-person perspective.

This epistemic guidance which we seek, which we need in order to proceed in our moments of doubt about what to believe, is arguably the very subject matter of a theory of knowledge. A theory of knowledge would provide the guidance we need, not by allowing us to unthinkingly defer to the instructions of some famous and long-dead philosopher, let alone some not-so-famous and still-living philosopher, but by providing a series of instructions which we could think through for ourselves and, if they are good instructions, ones we could recognize as good from our own perspective. This is just to say that the ultimate decisions here, the ultimate test of any guidance we may consider, must come from the first-person perspective. This is what Descartes attempted to provide when he wrote the *Meditations*, and when he wrote the *Rules for the Direction of the Mind*, and this is what countless epistemologists since Descartes have attempted to do even when they disagreed with him about what those rules might be. This ensures the importance of the first-person perspective.

We will begin this chapter with a description of the process of deliberation about what to believe as seen from the first-person perspective, that is, a description of how such deliberation appears to the person doing the deliberating. One might think that this is all that an examination of the first-person perspective could amount to. What else, after all, is there left to be said about this perspective?

There is, however, a great deal more to be said. The first-person perspective is just one perspective on what happens during deliberation. We can, in addition, take a third-person perspective on deliberation. This amounts to looking at the first-person perspective itself from the outside. While the deliberator has a unique perspective on his or her own deliberation, social psychologists have examined what goes on in us in these very acts of thinking about what to believe. Their experimental work offers a strikingly different account of the deliberative process. We will look at this work in detail and try to make some sense of how to reconcile these differing perspectives on what goes on in us when we deliberate. We cannot allow the first-person perspective to have the last word on the deliberative process. We cannot even allow the first-person perspective to have the last word on the first-person perspective itself.

4.2. Deliberation from the Perspective of the Deliberator

We need to look at two different kinds of situation in which we deliberate about our own beliefs: the situation in which we already have a belief on some matter, but stop to reconsider whether we really ought to believe as we do; and the situation in which we are uncertain what to think about some matter, and stop to deliberate about what we ought to believe. It will be useful to consider a typical case of each kind. As befits an examination of the first-person perspective on belief acquisition and revision, I will describe these cases in the first-person. In reading these cases, you should imagine yourself in the same situations, deliberating in just the ways I attribute to myself.

My daughter is planning to buy a used car, and having made very few such purchases before, and knowing that I have bought quite a few cars over the course of my life, she asks for my advice. She shows me some information about a car she is thinking of buying, and she asks for my opinion: Is this a good car for the money? I look at the information she has, and I'm familiar with cars of this sort. I have an opinion on this. Of course, one needs to check things out further—have a mechanic look the car over and so on—but this looks like a good car at a good price. I could tell my daughter what I think right away, but I don't do that. This is too important. So I stop and think again. I go over the information she has provided once more. I remind myself of what I know about cars of this make and model. I mentally rehearse both the good and the bad features of the potential purchase, since there are both good and bad features, and then I reconsider my judgment about the advisability of the purchase. Does my careful scrutiny of the evidence, evidence I already had and on the basis of which I had already formed a belief, genuinely support the belief I already held, or do I need to revise my belief in light of this more careful examination of the evidence? That is what I ask myself before I respond to my daughter's request for advice.

The reason I stop to scrutinize the evidence, to think my way through it and self-consciously question just what it shows, is that I believe that in doing this I will be more likely to arrive at an accurate opinion about the car than if I just accepted the belief that was produced within me immediately on being asked my opinion. I'm well aware that I have made mistakes in the past; that I have failed, at times, to give adequate weight to some of the considerations I should have taken into account; that I have sometimes looked over various fact sheets too quickly and missed things that a slower and more deliberate approach would have revealed. So my motivation in slowly and deliberately reviewing my evidence and its significance is just that I am convinced that this will provide an extra check on my belief. If it checks out, then I should go on believing as I already do and report this belief to my daughter. If, however, careful scrutiny of the evidence leads me to believe that I have been too hasty in my opinion, that I should have believed otherwise, then I should revise my belief accordingly and let my daughter know just what this more thoughtful and responsible approach has revealed. In undertaking this reflective examination of my evidence, I believe that I am capable of changing my belief should that examination give me reason to do so, and, of course, that I am capable of continuing to believe as I already do if my re-examination of the evidence confirms my original belief. What I believe is up to me, and the whole point of my reflective self-scrutiny depends on my being able to make appropriate changes in my belief if they are warranted, and on leaving things as they are if, instead, that should be warranted. I believe I can do these things.

What happens when I do stop to self-consciously think things through? I ask myself just what counts in favor of this particular purchase, and one by one, I bring various bits of evidence to mind. Then I do the same with the considerations which count against the purchase. I think about how much weight should be given to each of these considerations, and, having brought each piece of evidence to mind and assigned it its proper weight, I arrive at a view about what, all things considered, my evidence supports. I adjust my belief accordingly, or, if no adjustment is required, I recommit to the belief I had before all this reflective examination of the evidence. Either way, I take myself to have a better-supported belief than I did before, and I know full well just what my current assessment is based on. After all, I just went through this careful review of the evidence.

This is how my deliberations proceed when I already have an opinion on the matter about which I am deliberating. How are things different when I stop to deliberate without having any such opinion in advance of deliberation?

My son calls me and tells me about the results of the latest polling of public opinion about the coming presidential election, and he asks me what I think it shows. There are times when I hear or read about such polls and I am immediately led to an opinion about what they show. In some cases, the upshot is, as people say, a no-brainer. But this case is different. It is not obvious what the polls show. I don't find myself with a quick answer to my son's question. I need to think about it.

So I stop to reflect on the polling data. Once again my motivation is to figure out what the evidence shows, and, thereby, what I ought to believe. I bring to mind each bit of evidence my son has told me about, and I remind myself, as I do this, what I know about polling results, what I know about previous polls and how these new results differ from earlier ones, and a variety of other things which are relevant to assessing the upshot of this new information. I do this because, without such a careful reckoning, I just don't have any belief about the matter my son is asking about. By reflectively reviewing the evidence, I believe that I can arrive at a well-supported opinion. I believe that once I figure out what the evidence supports, assuming, of course, that I don't find that this is just too complicated for me, I can form a belief that is appropriately responsive to the evidence. And once I do form such a belief, I am well aware of the basis on which I formed it. After all, I just self-consciously reviewed the relevant evidence, and formed the belief on the basis of that reflective examination.

This is how deliberation about what to believe appears from the first-person point of view. In both of these cases, the motivation for undertaking such deliberation is to arrive at a well-supported belief. When I have an opinion even before undertaking any self-conscious deliberation, I believe that the act of deliberating will make me more reliable than I would have been without it. If I didn't believe this, there would have been no point in stopping to deliberate. And in the case where I have no antecedent belief, I believe that if I do form a belief as a result of deliberating, it will be well supported by the evidence. If I didn't believe this, then here too, there would be no point in stopping to deliberate. In both cases, I believe that what I come to believe is up to me, that I have control over what I believe. And in both cases, I believe that the act of deliberation puts me in a position to know just what it is that my new opinion is based upon. This is how deliberation appears to the person doing the deliberating.

4.3. Some Factors Involved in Reflective
Checking on Beliefs One Already Holds

Richard Nisbett and Timothy Wilson performed a landmark series
of experiments which they reported on in their 1977. In one of these
experiments, people at a shopping mall were asked to examine four
nightgowns and to report which of them they believed to be the best.
The experiment was set up so that the four nightgowns were as identical as such consumer goods can ever be. Most of those who were asked
their opinion said that they thought they were all about the same. But
Nisbett and Wilson put the question to them again: yes, they said, they
are all quite similar, but look them over carefully and tell us which you
think is best. In such a forced choice situation, one would expect that
those who agreed to answer the question would be more or less evenly
divided among the four nightgowns. Roughly one-quarter of those
asked should have chosen the nightgown furthest to the right; another
quarter should have chosen the nightgown that was second from the
right; and so on. But this is not what happened. Nisbett and Wilson
found that the nightgown furthest to the right was chosen far more than
any other. They concluded that subjects have a preference, other things
being equal, for objects further right than others. This is, to be sure, a
strange preference to have, but Nisbett and Wilson did not believe that
subjects were aware of having this preference. It's just that it seemed that
nothing but such a preference could explain the experimental results.

Now some have disputed that there is, in fact, such a rightmost preference. It could have been, for example, that subjects had a preference,
instead, other things being equal, for whichever nightgown was examined last, and that most subjects examined the different nightgowns by
starting on the left and working rightward. It wasn't being on the right
that really mattered; it just so happened that being on the right coin-
cided with being the last examined. Perhaps that is so. But which of
these two preferences was at work, or any other that would account for
the result, doesn't really matter. It is not disputed that subjects in situations like this tend to choose the object which is in fact furthest right,
whether they choose it because it is furthest right or for some other
reason. And, of course, such a preference is not remotely rational, on
whatever basis it is made.

The real issue for Nisbett and Wilson was not so much whether subjects came to believe that the rightmost nightgown was best because it was on the right or, instead, for some other reason. The real issue had to do with the explanation subjects gave for the preferences they had. None of the subjects suggested that they chose the nightgown they did because it was on the right, nor did any suggest that it was chosen because it was the last examined. Instead, subjects explained that, after careful examination, they found that the nightgown they chose had better stitching, or was made of softer material, or some other reason which, if only it were true, would make the nightgown they chose genuinely better. But, of course, all of these reasons cited were based on false beliefs, since the nightgowns did not differ in any respect at all. Subjects believed the claims they made. They genuinely thought that the nightgown they chose had better stitching, or softer material, and so on. And they genuinely believed that they had made their choice on the basis of this feature. But the nightgowns they chose did not have the features they attributed to them, and so they obviously could not have chosen them on that basis. These subjects were mistaken about the source of their beliefs about which of the nightgowns was best. They actually preferred the nightgown they did because it was on the right (or last examined, or some such thing); but they believed that they had made this choice, and that they formed the belief that their preferred nightgown was the best, on some other basis.

We are all susceptible, at times, to irrational influences of various sorts. We may be influenced in our choice of various consumer goods by their packaging, without realizing that this is influencing our choice. We may have a preference for one political candidate over another because that candidate reminds us of a friend or personal acquaintance we have strong feelings about, without realizing that this is the basis for our preference. We may find that a house we are considering buying is particularly attractive to us when the smell of chocolate chip cookies was in the air as we walked through its rooms, and yet be completely unaware of this influence. And we have all had the experience of misreading someone's intentions or attitude, judging them unfavorably, simply as a result of having had a difficult day. In these cases too, we do not recognize the source of our belief, and mistakenly attribute it to something else. Errors of these kinds are familiar to all of us; they are a part of everyday life. They are also well documented in the experimental literature. The Nisbett and Wilson experiments are a particularly vivid example of this.

The fact that we are systematically mistaken in our views about the source of our beliefs on these occasions is quite striking. It's not just that we fail to recognize the influence of certain factors on our beliefs. After all, we are often ignorant of many of the factors which influence the beliefs we hold. (Our judgments about the relative distance of objects is often influenced by the relative length of the shadows they cast. Who would have guessed that?) Rather, what is striking is that people in these situations have quite confident judgments about the source of their beliefs, and these judgments are systematically mistaken. Many of the subjects in Nisbett and Wilson's experiments confidently believed that they had chosen the rightmost nightgown as best because they noticed that it was made of finer material. More than this, their judgment about the source of their own belief was arrived at in just the way that such judgments are typically arrived at: they reflected on why they believed as they did, and it was just obvious! From the first-person point of view, from the perspective of the people who were making these judgments, it seemed that they could just tell what the source of their belief was by introspection.

How could introspective judgment about the source of these beliefs be so badly mistaken? When we stop to reflect, can't we just see, in some sense, where our beliefs come from? It certainly seems so from the first-person perspective. We regularly form completely confident judgments about the source of our beliefs. We just know, we may insist, why we formed the belief we did only a moment ago. But all of these cases, and many others like them, well documented in a wide variety of experiments, show that we are systematically mistaken in these confident judgments about the source of our beliefs when those beliefs are formed on an irrational basis.

Nisbett and Wilson offered an explanation for how this might come about. They argued that what seemed to these subjects like direct introspective access to the source of their beliefs was, in fact, the product of subconscious inference. Without realizing it, subjects were subconsciously reconstructing what the source of their beliefs must have been. They were, in effect, attempting to explain where their beliefs could have come from, and the resulting hypotheses they came up with seemed to them to be, not the product of subconscious inference, since they were, inevitably unaware of such subconscious processes, but, instead, direct awareness of the very source of the beliefs they had.

We might reconstruct the subconscious reasoning as follows. We all believe, of ourselves, that we are rational. I'm a rational person! And this belief is itself not an unreasonable one for each of us to have. But this view we have of ourselves might play a crucial role in our subconscious attempt to reconstruct what the source of our beliefs must have been. If I find that I believe that the nightgown on the right is the best of the four I've just examined, and I'm a rational person, then I must have noticed some feature of the nightgown which made it superior to the others. I wouldn't have chosen it on the basis of some feature which made it no better than the others! Well, what features make one nightgown better than another? Perhaps the stitching, or the material. That must have been it! If our attempt to reconstruct what the basis for our belief must have been itself presupposes that we reached that judgment on some rational basis, then it is unsurprising that the subjects in these experiments all attributed their beliefs to their noticing features of the nightgowns which, if only they had them, would have made the nightgowns better. Nisbett and Wilson were not suggesting that subjects self-consciously went through some reasoning like this. Rather, they suggested that reasoning like this went on in the subjects subconsciously, and thus without their being aware of it. It produced in them a belief about the source of their judgment that a particular nightgown was the best, and that belief about the source of their judgment seemed to be a product of direct introspection. This process of subconscious reconstruction which then appears to those who undergo it to be nothing more than direct introspection is known as confabulation.

Confabulation need not result in false beliefs about the source of our judgments. Indeed, it may well be that confabulation typically produces true beliefs, or at least beliefs which are approximately true. Most of our beliefs are not irrationally formed. The situation in which we judge the quality of consumer goods on the basis of their location is not at all the typical case, nor would subjects have shown the preference for the rightmost nightgown if the nightgowns had, instead, been quite different from one another. If one of the nightgowns were very well made, or made of clearly better material than the others, then subjects would have chosen that one as best whether it was on the right or not. The irrational factor which played a role in the Nisbett and Wilson experiments came into play only because there was no other basis for choice and subjects were forced to make a choice. But in the typical case, where there is some clear rational basis available for a choice, subjects will, in most cases, make their choice on that basis. And when they try to reconstruct, subconsciously, what their basis for choice must have been, their assumption that they made the choice on a rational basis will, in fact, be correct, unlike the subjects in the Nisbett and Wilson experiments. The hypotheses subjects come up with about what the basis for their choices and beliefs must have been in typical cases may, as a result, quite frequently be correct. So confabulation may actually produce true beliefs about the source of our judgments in most cases, even if, when we are subject to irrational influences, we will almost invariably come to mistaken views about the source of our judgments.

Although confabulation may thus typically produce true beliefs about the sources of our judgments precisely because our judgments are typically rationally formed, this should not be a great comfort to us. Remember that we stop to reflect on our beliefs in order to make sure that we have arrived at them in a way which is likely to lead us to the truth. We wish to weed out beliefs formed too hastily, too haphazardly, too casually, and so we stop to reflect on our reasons for belief. Our reflective scrutiny of those reasons is thus motivated by a desire to perform a bit of checking on our belief-acquisition processes, making sure that they were operating reliably. The problem, however, is that if our reflective checking procedure works in the way just described, then it is worse than useless in detecting our errors and shortcomings. Because our subconscious reconstruction of what our reasons must have been assumes that we arrived at our beliefs rationally, the result of our reflective check on our beliefs will do nothing more than reassure us that we were reasoning just as we should have, even when our beliefs were, in fact, arrived at unreliably. Instead of helping us to locate our errors, the process of reflective checking on our beliefs produces greater confidence that we were right, even when the initial process of belief acquisition was especially unreliable. The reflective process thus results in greater confidence in our beliefs without producing the greater reliability that we sought. When we stop to reflect, we are handing over the checking procedure to processes which systematically hide our errors from us, rather than reveal them.

There are other features of the processes which go on in us when we stop to reflect which come into play in hiding our errors from us. One of them has to do with the way in which memory operates. When my daughter asks me about the advisability of purchasing a certain car, and immediately upon reading the details of the car she is considering, I have a view about whether this is a car worth pursuing, there is an extraordinarily large amount of information I have stored in memory which bears on the advisability of this purchase. I may have owned cars of just this sort before. I will certainly have had friends or acquaintances with firsthand experience of this sort of car. I have read various consumer magazines about cars like this on the many occasions in the past when my wife and I were thinking about purchasing a car for ourselves. And I have had a good deal of experience with new car dealers, used car dealers, and private individuals selling their used cars. All of this information is stored in my memory and is relevant to the question my daughter is asking me. Without having to stop to reflect, I have already formed an opinion in answer to that question. If my automatic processes are working as they should, if they are working reliably, each bit of information which I have stored in memory was given its due weight when I formed an opinion without my having to self-consciously reflect on them one by one.

When I do stop to reflect, in order to check on the operation of those automatic processes, I make an effort to bring to mind all of that evidence which I have accumulated over many years. It can't all be brought to mind at once because the capacity of working memory, the region of the mind where self-conscious consideration of such matters occurs, is far too small to accommodate the vast amount of information I have stored in long-term memory, the repository of all I know and believe. I can, however, bring to mind, bit by bit, a good deal of information which bears on the issue under consideration. Some of the information I have stored in long-term memory is easily brought to mind. If I only recently bought a car just like the one my daughter is considering, a good deal of information about that car will immediately occur to me without any effort. If I had such a car very long ago, it may require some effortful attention to bring information about it to mind. When it comes to the availability of information stored in long-term memory, recency is one of the things that matters.

Here is something else that matters. It turns out that when we have a belief about some issue, information stored in long-term memory which accords with that belief is more easily brought to mind; information which runs counter to it is not so readily available. (See, e.g., Nisbett and Ross 1980, 180–83.) As a result, when we stop to scrutinize our evidence on some question about which we have already reached an opinion in order to make sure that we have properly accounted for all of our evidence, the evidence we bring to mind will not include all of our evidence, nor will it be a representative sample of the total evidence we have stored in long-term memory. Instead, the information which comes most readily to mind will be a deeply biased sample of our total evidence, strongly favoring the opinion we have already formed. This is one aspect of what psychologists refer to as confirmation bias or my-side bias. The way in which memory works thus biases reflective checking in favor of whatever we already believe. When we attempt to weigh the evidence for and against our antecedently held beliefs, the processes which allow us to bring that evidence to mind are placing a heavy thumb upon the scale. These processes are rigged in favor of the beliefs we already hold. No wonder that when we stop to reflect on whether the beliefs we hold are adequately supported by our evidence, we so often come to conclude that yes, indeed, we were right all along.

There is one further aspect of our reflective checking procedures which results in confirmation bias. Lord, Ross, and Lepper 1979 recruited two groups of experimental subjects. One group believed that capital punishment has a substantial deterrent effect: a policy of enforcing a death penalty for certain crimes makes it less likely that those crimes will be committed. A second group believed that capital punishment has no such deterrent effect. Subjects were then given two different studies to read. One provided evidence that capital punishment does have deterrent effects; the other provided evidence that it does not. After reading the two studies, subjects were again asked their opinions on the matter. Both before and after reading the two studies, subjects were also asked to rate their degree of confidence in their opinion.

One might have expected, or one might have hoped, that providing subjects with mixed evidence of this sort would serve to moderate the strength of the subjects' opinions. After all, those who believed that capital punishment has deterrent effects were now in possession of some evidence that it does not. Similarly, those who believed that capital punishment does not have deterrent effects were now in possession of some evidence that it, in fact, does. At a minimum, one might have thought that this new evidence would leave things as they were, since evidence was presented on both sides of the issue. But neither of these things happened. Instead, subjects tended to go on believing as they had before they saw the two studies, but they became even more confident that they were right than they had been before they saw the new information. Far from moderating the strength of the disagreement between the two groups, providing them with mixed evidence served to push them further apart.

The reason for this was not hard to find. When asked about the basis for their opinions after reading the two studies, subjects tended to find the studies which conflicted with their antecedently held beliefs to be less convincing than those which supported those beliefs. They were able to cite various methodological problems which these studies failed to deal with adequately. For example, one study compared two different states, one of which had a death penalty statute for murder and one of which did not. The state with the death penalty statute had a lower murder rate than the state which did not. Those who believed, before reading this study, that the death penalty does not have deterrent effects, pointed out, reasonably enough, that the two states differed in many other respects, and the difference in the murder rate might have been due to any of these other factors rather than to the fact that one state had the death penalty and the other did not. As a result, these subjects discounted this particular study. They found the second study they read, which provided evidence that the death penalty does not have deterrent effects, to be carefully and responsibly carried out. Subjects who antecedently believed that the death penalty does have deterrent effects found this second study unconvincing, and had quite reasonable methodological objections to it. They therefore discounted the study which seemed to show them mistaken, and embraced the study which supported their view as further evidence that they were right all along.

What we can see, looking at both sets of subjects, is that they all treated the new evidence less than evenhandedly. When they were asked to reflect on what this evidence showed, evidence in favor of their prior belief was found to be convincing; evidence against that belief was found to be unconvincing. Given that, they each increased their confidence in their prior belief. Carefully reflecting on the value of new evidence did not make these subjects more reliable, it only made them more confident in what they had believed from the beginning.

These experiments do not show that reflecting on one's reasons for belief, or reflecting on new evidence, can never make one more reliable. What they do show, however, is that when one stops to reflect on these matters in a situation in which one already has a belief, there is a very strong tendency to find that one's evidence does, indeed, provide good support for the belief one has, and thus a very strong tendency to go on believing as one did before. This conservative tendency is not moderated when the reasons for which one actually believes something do not provide such support. When one's belief was the product of irrational or nonrational influences, one is still likely to come to believe, when one reflects on the quality of one's evidence, that one has very good reason to go on believing as one did.

The checking procedures which come into play when we stop to reflect, when we are motivated to make especially sure that we are believing as we should, are ill-suited to that task. Rather than locate our errors and improve our reliability, these procedures are likely to hide our errors from us and make us more confident of the things we already believed, whether we had good reasons for those beliefs or not. The first-person perspective, the perspective of the person who deliberates about what to believe, in situations where that person is checking on an already existing belief, is systematically misleading.

4.4. Reflection on What to Believe in the Absence of Preexisting Belief

Beliefs already held bias reflective checking. But this is not the only situation in which we reflect on our reasons for belief. We also stop to reflect on our reasons in situations where we have yet to form a belief. In these situations, there is no preexisting belief to bias our reflective self-examination. In these situations, as we have noted, we may carefully reflect on our evidence, attempting to figure out what belief the evidence supports. Although we sometimes find, even after such reflection, that we still don't know what to believe, our reflective activity does often reach a successful resolution: we come to regard our evidence as providing us with good reason for some particular belief. More than this, when we are in such a situation, and when we do form a belief as a result of our reflective activity, we find ourselves quite confident that we know why it is that we formed the belief we did. We know the reasons for which we arrived at the belief because we just formed the belief as a result of self-conscious scrutiny of those reasons. Or so it seems to the reflective reasoner. But is the reflective reasoner right about this? Does the first-person perspective on belief acquisition under conditions of reflection give an accurate view of the process of belief acquisition, at least when the reflective reasoner does not already have a belief on the matter under consideration before starting to reflect?

I want to divide this question into two parts. First, does the perspective of the reflective reasoner give a complete picture of the reasons for which the new belief is adopted? When the reflective reasoner comes to believe that they know what the reasons are for which they have formed their new belief, does their view of the matter leave anything out? And second, even if, the perspective of the reflective reasoner is only partial, that is, even if it does leave out some of the reasons for which their belief was formed, is their view of their reasons for belief accurate as far as it goes?

Let us go back to the case I described in section 4.2 in which my son asks me what I think the latest public opinion poll shows about the upcoming presidential election, and I don't immediately have a view about this; the matter is complicated, and I need to self-consciously think through the implications of this new evidence. In such a situation, I stop to reflect on the new evidence my son has given me, and I bring to mind a variety of considerations which help me evaluate its significance. I remind myself of what earlier polls showed so that I can see whether there is some trend in the direction of one candidate or another. I think about the reliability of this particular pollster, since some polling organizations have a very good track record, and others are less reliable. I think about the reliability of polling in general, and I remind myself about how even the most reliable pollsters make mistakes. And I think about issues having to do with the margin of error, the fact that any poll result should not be regarded as a very precise number, but rather as providing us with a good estimate, at best, plus or minus a certain amount. I self-consciously think about all these matters because I've thought about political polling a great deal, and only then do I come to an opinion. When I do come to an opinion, I can, it seems, confidently report exactly what the basis for my opinion was.

Should I be confident that my view of the reasons for which I formed my belief leave nothing out? It should be quite clear that I should not. The range of considerations which bear on the question I am asking myself, about what this new polling data show, is immense; after all, I've followed presidential elections, and other elections, for many years, watching the poll numbers change over time, and seeing some polls prove amazingly accurate and others prove horribly mistaken. I've been overconfident in my own predictions at times, giving some polls more weight than I should have, or giving others less weight than they deserved. All of this data about the polls themselves, and my own success or failure in being responsive to polling data, is relevant to the question my son asks me. And a huge amount of information on these matters is stored in my long-term memory. I don't, and can't, bring it all to mind. And even if I could bring it all to mind for conscious consideration, I couldn't possibly scrutinize all of this data self-consciously in a reasonable amount of time. This does not mean, however, that it is not playing a role in informing my reasoning. Some of it quite clearly does play such a role.

Consider a simple case. There are two people on the presidential ballot and my son has just told me what the polls show about the head-to-head numbers when likely voters are asked which of these two they expect to vote for. The contest between the two candidates has been in the news for months. I know, and every minimally aware voter knows, that these are the only two candidates on the ballot. In such a situation, I am very unlikely to bring to mind that there are no other candidates. I don't need to think about that; I take it for granted. The fact that these are the only two candidates informs the way I think about this issue. If there were some third candidate on the ballot, even a candidate who is not likely to receive a large share of the votes, then I would think about the upshot of this poll, which asked only about the two major candidates, in a completely different way. I don't have to bring to mind the fact that, in this particular election, these are the only two candidates on the ballot in order for my knowledge of that fact to influence the way I think about the matter. But what passes through my conscious mind, what I self-consciously consider when I think about the upshot of the poll, does not include the thought that these are the only two candidates. The first-person perspective on one's reasons for belief is inevitably partial. It leaves things out.

Consider another case. I'm playing chess, and it's my turn.[4] What move I should make is not completely obvious. I look over the board carefully, and two moves look promising. As a result, I think carefully about how each of these moves might play out: what my opponent might do in response to each, and whether that would leave me in a good position or not. I'm not a very good chess player, and I can't think very far ahead, but I do think about the implications of these two promising options.

Of course, there are many more than two different legal moves I might make, even though I only reflect on the upshot of these two. There are, in addition, countless ways in which the chess pieces could be moved on the board which would violate the rules, and I don't bring any of those possibilities to mind, even to dismiss them as unacceptable. I don't think about the various moves which the rules prohibit, one by one, eliminating each as a violation of the rules. I've played the game long enough, even though I'm not very good at it, so that I don't need to mentally rehearse how each of the pieces is permitted to move. In spite of this, it's quite clear that my knowledge of the rules governing the movement of each piece plays a role in how I think about what I should do next. It has to. It's not just a coincidence that I only consider legal moves. This, by itself, shows that what passes through my conscious mind when I consider what move to make offers me only a partial view of the reasons I have for the decision I ultimately make.

The same is true of the two moves I self-consciously consider. When I look at the board, and these two possibilities jump out at me as worthy of further consideration, I don't bring to mind each of the other legal moves available to me. There are many more than two ways I might legally move my pieces, but I don't stop to reflect on each of them, even to quickly dismiss all but two as unpromising. I just immediately fasten on two promising possibilities in this case as being the ones worthy of further consideration.

How did I fasten on these two as worthy of further thought? I just did, it seems. My mind was immediately drawn to them, and I then moved on to think about how my opponent would respond were I to make each of these moves. I didn't stop to think about other possibilities; I didn't need to. It's quite clear, however, that the selection of these two possibilities was not merely random. It couldn't have been, for, as we have seen, it's not just a coincidence that both of the moves I consider are legal moves. Even among the legal moves, however, it can't just be a coincidence that I focus on these two. If one were to choose randomly among all legal moves, one would play a far worse game than I do. Most of the legal moves are just awful, and although I'm not very good at chess, I'm not so bad that my moves are no better than ones randomly chosen among the legal possibilities. What this means, then, is that even though I do not self-consciously consider more than two possibilities, I must have had my reasons for choosing these two moves to think about further rather than any of the others. These reasons did not pass through my conscious mind, but they must have been guiding my choices without my being aware of their influence. Any proper explanation of why I moved as I did, and why I considered the possibilities I did, would need to explain why I thought about the two good possibilities, and why I didn't stop to consider the others. Not all of the reasons for my choice, and not all of the reasons for my belief about which moves are best, are present to consciousness. The first-person perspective on my reasoning leaves out much of the reasoning that goes on within me. The first-person perspective on reasoning is inevitably only partial at best.

Even if my perspective on my own reasoning will inevitably leave out many of the reasons on the basis of which I form the beliefs I do, does it at least give an accurate picture of my reasoning as far as it goes? When I self-consciously consider a variety of reasons for forming a belief and self-consciously decide that these are, indeed, my reasons for taking on that belief, must I at least be right that these reasons I take to be my reasons are, in fact, among the reasons for which I form my belief? Here, too, the first-person perspective need not represent our own mental processes as they truly are.

Let us return, once again, to my son's question about the latest public opinion polls on the coming presidential election. When I reflect on the significance of these polls, I may come to believe that they provide good reason to believe that a particular candidate is currently ahead in the race. On that basis, it may seem to me, I form the belief that that candidate is, in fact, ahead, and I tell my son that this is what I believe, and I tell him why it is that I believe it. Must I be right about this? It seems not.

A friend of mine, a philosopher who need not be named here, has pointed out that my views about political outcomes over the years have been remarkably optimistic. This friend has noted that when I favor some outcome, I tend to believe that it is more likely to occur than the available facts would justify. When I fear a particular outcome, this same friend has noted, I tend to view it as less likely to occur than a clear-eyed view of the facts would allow. My beliefs, he has pointed out, are driven, in these situations, by my hopes and fears far more than by the evidence.

Perhaps my friend is wrong about me. (I'd like to believe that!) I think we can all agree, however, that I cannot show that my friend is wrong by pointing out that when I thought about the latest polling, it didn't seem to me that I was driven by my hopes and fears. My friend is not suggesting that I self-consciously take these things into account. Rather, he is suggesting that my hopes and fears subconsciously influence me; they play a role in determining my opinion without my realizing it.

If I am influenced in this way, I am not the only one. There is now a very large body of work in social psychology on the phenomenon of implicit bias. It is well documented that people may be influenced in the judgments they make about other individuals because of their appearance, their race, their gender, their accent, and so on. Of course, some people form judgments about others knowing full well that these factors are influencing them; they purposely take these factors into account. The phenomenon of implicit bias, as the name suggests, is not about such cases. Rather, what is well documented is that these factors may play a role in the judgments people form even when they sincerely believe that they are uninfluenced by such factors.

It will be helpful to describe one of the studies from this body of work. Bertrand and Mullainathan 2004 took a résumé and printed two versions of it. In one version, the résumé had the name Emily Walsh printed at the top; the other résumé, otherwise identical to the first, had the name Lakisha Washington at the top. These names were chosen because of the ethnic identities these names conjure up in the minds of most people who see them. The résumés were then sent out to a very large number of job listings posted in Boston and Chicago newspapers. "Emily" got 50% more callbacks for interviews than "Lakisha."

This result could be attributed, perhaps, to nothing but fully intentional bias. Perhaps all of those who responded more sympathetically to "Emily's" résumé than "Lakisha's" were out and out racists, thinking to themselves all the while that they didn't want to employ anyone named Lakisha. No doubt this did account for some of the result. But to suggest that this result was exclusively the product of biases of which the employers were fully self-aware would be overly optimistic. The result is, in some ways, far more concerning than that.

Project Implicit, which has investigated the phenomenon of implicit bias in great detail, puts this study in broader perspective. Just about everyone is familiar with a variety of stereotypes of various racial and ethnic groups, of different genders and groups with different sexual preferences. Some of these stereotypes may include positive features and features which are neutral, but many of them include quite negative features, portraying one or another group in extremely disparaging ways. One may, of course, be familiar with such disparaging stereotypes while recognizing, at the same time, that they are inaccurate. Every reader of this book, I expect, can list a variety of such disparaging stereotypes which they know full well to be inaccurate. Many of us, for example, are familiar with a stereotype of our own ethnic, racial, or other social group—how could we fail to be aware of this?—while recognizing just how inaccurate, and unjust, that stereotype is. What numerous studies have revealed, however, is that merely being acquainted with a stereotype, even when one knows it to be inaccurate, will often have an influence on our behavior and the judgments we make. This influence operates subconsciously; we are not aware of it;

we certainly don't explicitly and self-consciously reason from stereo-types we reject to conclusions they would only license were they accu-rate. Nevertheless, knowledge of these stereotypes plays a role in the way we act and the way we think about the world. Explicit prejudice is a horrible thing, but the influence of implicit bias shows that even those who sincerely reject inaccurate and unjust characterizations of various groups are not immune to the effects of the stereotypes they themselves reject.

This will, to many, sound just unbelievable. "I'm not like that!" they will insist. "Maybe other people are influenced in this way, but I'm more thoughtful, more educated, more decent than the people who unthinkingly respond in this way!" This is, unfortunately, a common reaction to these studies. That is why you should put down this book right now and take one or more of the implicit association tests. You can find them at https://implicit.harvard.edu/implicit/. Do it now. It will only take you a few minutes. No amount of argumentation on my part, no careful presentation of statistical results, no detailed discus-sion of experimental methodology will allow you to appreciate the im-portance of this phenomenon in the way that simply taking these tests yourself, and seeing the influence of a knowledge of common stereo-types on your own behavior and judgment, will allow. Do it now. The rest of this chapter can wait.

This work shows the influence of our knowledge of stereotypes on our behavior and beliefs in ways of which we are unaware, and which a first-person examination of our own thinking will not reveal to us. Even this, however, minimizes the significance of these results. The work of Project Implicit, and a good deal of related work in social psy-chology which has nothing to do with racial or ethnic biases, does not just show that when we stop to reflect on what to believe we will be influenced by things of which we are unaware. We have already seen, even in the very simple cases involving my beliefs about presidential polling or my belief about the best move to make in a chess game, that what passes through my conscious mind cannot possibly include all of the factors which play a role in producing my beliefs, or all of the reasons I have for forming the belief I come to adopt.

Although the work on implicit bias is of great social consequence, it might seem, at least on this point, to be just another illustration of the same phenomenon. The first-person perspective does not give us a complete picture of our reasons for belief.

Implicit bias also shows, however, that the first-person perspective on belief acquisition is often inaccurate, even as far as it goes. When a prospective employer reads a résumé with the name Lakisha Washington at the top and concludes that she is not a good candidate for the job, the employer will be able to cite reasons for coming to that conclusion. Thinking about whether to invite her in for an interview, he will be able to pick out various items on the résumé which justify failing to interview her, and he will believe that he reached the judgment about her unsuitability for the job solely on the basis of those very features of the résumé. He has just read the résumé; he noticed those features of her employment history; he knows why he decided to reject her application, or so he will think to himself. When the same employer reads an identical résumé with the name Emily Walsh at the top, he may find her a good candidate for the job, and he will be able to cite features of her background, right there on the résumé, which justify that conclusion. He will sincerely believe that he knows just why he came to the conclusion he did. But the fact that these two résumés are treated so differently not only shows that the employer is influenced by factors of which he is unaware. It also shows that the factors which he took to influence his judgment were not playing the role which he thought they did. The first-person perspective on our belief acquisition, when we deliberate about what to believe, even when we do not antecedently hold some opinion on the matter, is not only incomplete; it is often inaccurate as well.

It is important not to exaggerate these results. The effects of implicit bias are not so powerful that they will swamp all other evidence in every case; they won't. An absolutely extraordinary résumé, one which shows a person who is superbly qualified for the job, will not automatically be rejected just because the name at the top elicits some negative associations. Similarly, a résumé which shows someone demonstrably unqualified for a job will not automatically result in a favorable decision by the employer when the name at the top elicits some positive associations. Implicit bias does not determine every result in these cases. It does, however, influence judgments in that very wide range of cases in the middle of the pack, those résumés which have a mix of positive and negative features. In these cases, we are influenced by implicit associations, and we are ignorant of that influence. We believe our judgments on these cases are a product of the features of the résumés which we cite when asked to justify our judgments; but we are mistaken in thinking that our judgments are a product of those features.

Where does this leave us? Ziva Kunda provides a clear summary of the literature in social psychology which bears on this issue.

> Our judgments, feelings, and behaviors can be influenced by factors that we have never been aware of and have only been exposed to subliminally, by factors that we were aware of at one time but can no longer recall, and factors that we can still recall but whose influence we are unaware of. (1999, 308)

The first-person perspective on belief acquisition, the perspective we take on our own belief acquisition when we stop to reflect, gives a vivid and psychologically compelling view of the basis on which we form our beliefs. This view is neither complete nor always accurate as far as it goes. When we are most in need of correction, when our beliefs have been formed in ways which are unreliable, when those beliefs are responsive to things we should not take into account, and when they are insufficiently responsive to things which we should take into account, the first-person perspective will not typically reveal these shortcomings to us but will instead provide us with a view of ourselves as model epistemic citizens, forming our beliefs exclusively on the basis of good reasons. The first-person perspective gives us a flattering but inaccurate view of our own reasoning, and it does not typically aid us in locating our errors.

4.5. Conclusion

We sometimes stop to reflect on our beliefs because we want to be especially careful. We want to run an extra check on some belief to make sure that we are believing as we should. We review the reasons we have for holding the belief, and we review our evidence, for and against the belief we already hold, to see whether we should go on believing as we do, or, instead, make some revision in our view of the world. Our motivation on these occasions is clear: we want to check on our own reliability. Much like the person who wiggles the door handle to make sure that the door is locked when leaving the house, our reflective checking on our beliefs is undertaken with the thought that checking on a process we typically engage in unreflectively will help us discover any mistakes we might have made and thereby allow us to correct them. When we reflectively check on our beliefs in this way, we seem to have a clear view of our reasons for the belief under review, and seem to be able to undertake a clear-eyed assessment of the quality of our reasons.

This is how it seems from the first-person perspective, that is, from the perspective of the person who is re-evaluating their own belief. What actually goes on during this process of reflective checking, however, is not always what it seems. The perspective of the individual deliberating about the strength of their reasons for belief, however vivid and compelling they may seem, is only partial. It leaves out a very large number of factors which play a role in belief formation. It leaves out many of the reasons for which that person formed their belief. More importantly, the picture it provides of the process of belief acquisition is not only partial; it is often inaccurate as well. As a result of a number of factors which bias the process of reflective checking, the person who checks on the legitimacy of the beliefs they already hold is likely to come to the conclusion that they did, indeed, hold their beliefs for good reasons, whether they in fact hold their beliefs for good reasons or not. Given the ways in which the process of reflective checking is biased, and biased in ways that are invisible to the person doing that checking, the person who checks on their reasoning because they wish to be especially careful or responsible on some particular issue is likely to end up more confident in whatever they believed before they undertook that process of reflective checking, but they are not likely to uncover any errors they might have made. Reflective checking on our beliefs typically makes us more confident, but no more reliable. The first-person perspective on existing beliefs offers us a distorted view of our reasons for belief in just those cases where we most need correction.

We also engage in reflection on our reasons for belief in cases where automatic processes have simply failed to produce a belief on a matter that we wish to form an opinion about. We stop to reflect on the evidence about an issue in the desire to achieve some resolution of the matter as a result of such careful and systematic evaluation of the evidence. When we do come to a belief as a result of such self-conscious review of our evidence, we are, inevitably, confident that we know what the basis for our belief is. We have, after all, just thought through those reasons. Here too, however, the first-person perspective on our deliberations is systematically misleading. The considerations which we are able to bring to mind reveal only a small fraction of the considerations which actually influence us. The reasons we have for coming to a belief, even in this case, vastly outstrip the reasons we are able to bring to mind. This means that our view of our reasons for belief, even after careful reflection, provides only a very partial picture of the reasons which played a role in our coming to believe as a result of that reflection. And here too, the picture that self-conscious reflection on what to believe offers us is not only partial; it is also frequently inaccurate, giving us the illusion that we are moved by reasons which played no role in influencing what we came to believe, and the illusion that we were uninfluenced by reasons which did, in fact, play a crucial role in determining our belief. Our view of our own belief acquisition, when we reflect, is likely to be least accurate on just those occasions when our process of belief acquisition was most unreliable.

Suggestions for Further Reading

The history of epistemology is dominated by theorizing undertaken from the first-person perspective. Descartes is of paramount importance here. A very nice contemporary articulation and defense of this way of viewing matters may be found in Richard Foley's "What Am I to Believe?," in S. Wagner and R. Warner, eds., *Naturalism: A Critical Appraisal*, University of Notre Dame Press, 1993, 147–62.

Nisbett and Wilson's landmark paper discussed in this chapter is "Telling More than We Can Know: Verbal Reports on Mental Processes," *Psychological Review*, 84 (1977), 231–59. A very readable overview of the literature on the distortions which the first-person perspective gives rise to may be found in Timothy Wilson's book *Strangers to Ourselves: Discovering the Adaptive Unconscious*, Harvard University Press, 2002.

A very useful overview of the literature on implicit bias and its implications is the two-volume *Implicit Bias and Philosophy*, Michael Brownstein and Jennifer Saul, eds., Oxford University Press, 2016.

5

From the Individual to the Social

5.1. Beyond Individual Cognition

With the exception of the brief discussion in section 2.5, our entire focus thus far has been on individual cognition. There is nothing exceptional about this approach. The vast majority of philosophical discussion of the nature of knowledge has been focused on individual cognition. This is in part a byproduct of the fact that much of that philosophical discussion has proceeded from the first-person perspective, examining questions about how one ought to arrive at one's beliefs from the perspective of the would-be believer. In recent years, however, there has been growing attention to social factors in knowledge acquisition, and for good reason.[1]

There is no denying that cognitive processes which take place in individuals, such as perception and inference, play a central role in the acquisition of knowledge. We are, however, social animals, and the acquisition of knowledge is often the product of social activity. In the sciences, research is typically the product of groups of individuals working together on a common problem. A great deal of thought goes into the management of such groups so that the division of intellectual labor which group investigation embodies can most profitably be brought to bear on the questions the group is trying to answer. At a higher level, private, national, and international funding organizations give a good deal of thought to the division of intellectual labor, so that promising lines of inquiry are not neglected, even if any particular research group will, inevitably, have only a limited focus. Questions about how research should be organized, and how best to manage the division of intellectual labor, fall within the province of epistemological theorizing, and these are questions, not about individual cognition, but about the acquisition of knowledge as a product of group activity.

Scientific Epistemology. Hilary Kornblith, Oxford University Press. © Oxford University Press 2021.
DOI: 10.1093/oso/9780197609552.003.0005

Relevant here too are issues having to do with the dissemination of knowledge. Research groups need to communicate their results to the wider scientific community, and the ways in which conferences and scientific journals ought to be organized so as to facilitate the spread of knowledge, and to minimize the chances that misinformation is spread, are an important topic in social epistemology as well. Nor are these issues about the dissemination of knowledge limited to the spread of knowledge within the scientific community. Many scientific issues are of great interest and great import to the wider public, and questions about how best to disseminate such knowledge and how to minimize the spread of misinformation at this level are the focus of much current research in social epistemology.

In the sciences, these issues about the proper organization of individual research groups and the proper distribution of intellectual labor in the wider scientific community, as well as the various issues about the dissemination of knowledge, are often the focus of careful thought on the part of scientific investigators, laboratory managers, funding groups, scientific agencies, professional organizations, and journal editors. The social acquisition of knowledge is not limited, however, to the realm of science. We are all constantly interacting with others, and even the most informal conversations often serve as a medium for joint investigation of a question of common interest, or as a way of disseminating knowledge. If you and I chat about some new restaurant in town or a new movie, about travel or politics, about family or friends or finances, we may disseminate information and thereby aid in the transmission of knowledge, or we may disagree and I may seek to convince you that I have got it right. These social interactions, however informal, serve to inform, and, at times, to misinform; they serve to advance our knowledge and understanding or, at times, to impede knowledge. Just as the various processes by which beliefs are acquired and revised by individuals, in the absence of social interaction, are a proper focus of epistemological investigation, the social processes by which beliefs are acquired and revised are also a proper focus of such investigation.

This chapter will be concerned with a single example of the way in which social interactions may serve as a medium for the acquisition and revision of belief so as to illustrate the ways in which a social perspective may illuminate epistemological issues. The example I will focus on derives from work by Hugo Mercier and Dan Sperber (2017). The best way to approach this work is by starting with a puzzle.

5.2. A Puzzle about the Human Capacity to Reflect on Our Beliefs

Here are two striking facts about human beings. First, we seem to be the only animals with the capacity to reflect on our beliefs and to think about the reasons for which we hold them. And second, the human species, quite clearly, has by far the greatest intellectual achievements of any animal species on earth. Our scientific achievements alone are more than sufficient to make this clear.

It doesn't seem that it could be a coincidence that the species which is able to reflect on its beliefs and its reasons for belief should also be the species with the greatest intellectual achievements. Surely the capacity to reflect in this way must be instrumental in producing those achievements. Indeed, there is an obvious explanation for how the capacity to reflect could lead to such beneficial results. Human beings and many other animal species have impressive capacities for gaining perceptual knowledge. In addition, both human beings and many other animal species are natively endowed with an impressive array of other automatic cognitive processes which are instrumental in gaining substantial knowledge about their environments. These various automatic processes, as we discussed in Chapter 3, are quite reliable in normal environments, but they produce a characteristic pattern of errors when they operate outside those environments. Creatures endowed with processes like these would do well in the acquisition of a great deal of knowledge, even if, were they reliant on nothing but such processes, they would also acquire many false beliefs.

We human beings, however, are not reliant on these processes alone. We have the capacity to reflect on our beliefs, and on our reasons for belief, and this provides us with a degree of intellectual flexibility which other creatures do not have. There is thus an obvious explanation for why human beings, the only creatures who are able to reflect on their beliefs and their reasons for belief, should also be the creatures with the greatest intellectual achievements: when we engage in private moments of reflection, thinking about whether beliefs we have already formed are really ones we have good reason to hold, and thinking about whether beliefs we might take on are really ones we have good reason to adopt, these acts of reflection allow us to correct errors we would otherwise make if not for our capacity to reflect in these ways. The capacity for such private reflection, of just the sort that Descartes and so many later generations of epistemologists urged us to engage in, thereby allows for greater reliability than creatures who lack the capacity to reflect. The connection between our two striking facts about human beings seems straightforward.

As Mercier and Sperber point out, however, there is only one problem with this explanation. It is clearly false! As the work documented in Chapter 4 makes clear, private reflection on our beliefs and our reasons for belief is not typically effective in allowing us to detect our errors and thereby correct them. Instead, private reflection on such matters typically hides our errors from us and makes us more confident, but no more reliable than we were before we began to reflect. The proposed resolution of our puzzle is thus completely bogus. Private reflection does not typically allow us to achieve greater reliability than we would have without it, and it cannot, therefore, explain how we are able to achieve so much more knowledge than creatures who lack this capacity.

And so we have a puzzle. Must we think that our two striking facts about human beings are just a coincidence? That would be very hard to believe. There are a wide range of processes which go to work in allowing us to reflect on our beliefs and our reasons for belief. The capacity to reflect in this way is exceptionally complex. Where complex organs are found in various species, we should expect to find that these organs were selected for by evolutionary forces. It is not just a coincidence that we happen to have hearts and that they pump blood throughout the body. Hearts have a function—to circulate the blood—and the best evolutionary account of why it is that various species have hearts is that hearts were selected to perform this function: their very existence is explained by the fact that this is what hearts do. Just as the existence of complex organs is best explained by way of the function for which they are selected, the same is true of complex mental capacities. It is not just the sense organs themselves which require an adaptationist explanation; the mental capacity to process the information which the sense organs provide is also best explained in this way. Indeed, the sense organs would be of no use whatsoever without the mental capacity to draw on the information they provide. An adaptationist explanation of the existence of sense organs thus requires an adaptationist explanation of the capacity to process sensory information. Given the complexity of the various processes which jointly constitute our capacity to reflect on our beliefs and our reasons, we should thus expect to find an adaptationist explanation for the presence of such a capacity in the human species.

And that, of course, is precisely where the puzzle lies. The obvious adaptationist explanation, that these processes were selected for the contribution they make to greater reliability in belief acquisition and revision when we engage in private reflection, is not correct. This can't be the function of our capacity to reflect because that capacity is so manifestly ill-suited to that task. If that is not the function for which our capacity to reflect was selected, then what is the function of that capacity? What end does reflection serve?

5.3. An Adaptationist Hypothesis

Mercier and Sperber have an hypothesis. What they suggest is that the capacity to reflect was not selected for private reflection. Rather, the capacity to reflect was selected for use in social interactions. In particular, social creatures frequently engage in cooperative behavior, and the capacity to reflect on our reasons aids cooperation. If you and I have a shared goal, perhaps to hunt down some animal for dinner, we need to coordinate our plans to achieve that goal. You may have one idea about how best to track our target, while I have quite a different idea. If you can explain your reasons for doing things your way, and I can explain my reasons for following my plan instead, then we can try to figure out which is the better plan. In order to do this, we each need the capacity to reflect on our reasons so as to be able to articulate and defend them, and thereby resolve our differences. We might each, on our own, do perfectly well in engaging in successful solitary behaviors without that capacity. Such solitary projects require a capacity to develop plans which have some reasonable chance of success, but the success of such plans doesn't require the ability to explain, either to ourselves or to others, why anyone should think that these plans are good ones. As long as the plans are in fact good, we will be likely to succeed. When we attempt to cooperate with others, however, we need to be able jointly to settle on a plan, and since different individuals will often have different ideas about the best plan to follow, some way to resolve our differences is needed. The capacity to reflect on our reasons and thereby articulate them contributes to our ability to settle on a plan. And since there are many tasks, such as hunting down large game for dinner, which are far better pursued by a group than by an individual, there is a great advantage to any species that has such a capacity.[2]

When I try to convince you to follow my plan, I need to tell you not only what my plan is, but why I think it is a good plan to follow. You, of course, have a different idea, and you try to convince me that yours is the better plan. We each need to have certain intellectual capacities if this interchange is to facilitate successful cooperative activity: we need to be able to articulate reasons for thinking that our own plan is a good one; we need to be able to find faults in alternative plans if those plans do have faults; and we need to be able to evaluate the reasons which our would-be cooperative partner may offer.

Mercier and Sperber argue that the very features which make reflection work so badly when we engage in it privately actually serve to aid in cooperative ventures. Consider our earlier discussion in section 4.3 of confirmation bias or my-side bias. We pointed out that someone who reflects on a belief they hold in order to see whether they hold it for good reasons is likely to come to believe, as a result of reflecting on their reasons for belief, that they did have good reasons for holding that belief, and they are likely to think this whether they do, in fact, hold the belief for good reasons or not. This is why private reflection does not make us more reliable. It hides our errors from us by giving us the impression that our reasons for belief are good ones even when they aren't. We noted, as well, that evidence against beliefs we already hold is treated differently than evidence in favor of those beliefs. We are quick to find problems with apparent evidence against our beliefs, but subject evidence in favor of our beliefs to far less searching scrutiny.

And this too explains why private reflection does not typically contribute to greater reliability.

But now consider how that same process works when you and I are trying to figure out how to proceed in our cooperative venture. I propose a plan, and you suggest an alternative. We each need to offer reasons in favor of our proposed plan, and this is where reflection comes in. We each reflect on our reasons for favoring the plan we've proposed, and given how reflection works, we will each offer reasons which we ourselves find convincing. We won't be very good in spotting problems with our own reasoning because the way in which we reflect on our reasons is governed by confirmation bias. But I will be quite good in spotting difficulties with your plan, should there be any, and you will be quite good in spotting difficulties in my plan, should there be any. Confirmation bias, in this social interaction, serves to divide the intellectual labor: I do the work of spotting defects in your plan, something you will not be very good at; and you do the work of spotting defect in mine, defects which I will have been unlikely to notice. The net effect of such a dialogue, and such a division of intellectual labor, is that we will be able to converge on a judgment about which plan is better, and we will not only reach agreement on this, but the plan we ultimately agree on is likely, in fact, to be better than either of our original plans as a result of being subjected to assessment in this way.

Mercier and Sperber's adaptationist hypothesis, then, is that the capacity to reflect on our reasons was not selected for private reflection; it was selected for use in cooperative problem-solving. If they are right, the reason why reflection works so badly when conducted privately is that it was not adapted for such a use. Using reflection in private evaluation of the quality of one's own reasoning, on this view, is a bit like using a screwdriver to hammer in a nail. Screwdrivers weren't designed to do that, and although you can try to use them for that purpose, they won't work very well when conscripted for such use. Reflection was selected for use in cooperative social endeavors, and if one insists on putting it to use outside of those situations by reflecting privately rather than in the course of joint evaluation with others, it just won't work very well at all.

If Mercier and Sperber are right about this, then we have a solution to our puzzle. Reflection does improve reliability, so it is not just a coincidence that we are the only animals capable of reflection and also the animals with the greatest intellectual achievements. The way in which reflection leads to greater reliability, however, is not at all what we might have suspected. Greater reliability is not achieved by way of private reflection and private assessment of our reasons for belief. Contrary to the way it appears to those who engage in private reflection, contrary to the way it appears from the first-person perspective, private reflection does not help us spot our errors and thereby improve our ability to gain true beliefs. Reflection may instead be used in cooperative problem-solving with others. It is here that the manner in which we reflect, and the manner in which a group may jointly assess the reasons they have for belief, serve to improve the entire group's capacity to gain knowledge. Our capacity to reflect on our beliefs was selected for engagement in these cooperative social epistemic endeavors.

5.4. What Reason Is There to Think That This Hypothesis Is Correct?

There is no doubt that this is an interesting hypothesis about the function of our capacity to reflect on our beliefs and our reasons for belief. What reason do we have, however, to think that this hypothesis is correct?

The fact that this hypothesis, if true, solves our puzzle about the relationship between the human capacity for reflection and the impressive intellectual achievements of the human species is one thing that counts in its favor. We have seen that the obvious explanation for why it is that humans, who are the only animals capable of reflection, are also the ones with the greatest intellectual achievements—namely, that private reflection improves reliability—is mistaken. And it is also clear that it just doesn't seem that it could be a coincidence that great intellectual achievement and the capacity for reflection are found together. When an hypothesis explains an otherwise puzzling phenomenon, and no other explanation for that phenomenon is available, this, by itself, counts heavily in favor of the truth of that hypothesis.

We need to be careful here, however. We need to see exactly how reflection is supposed to operate in cooperative group problem-solving so as to improve group reliability despite the fact that private reflection does not improve reliability. The suggestion that reflection works in this manner may seem almost magical. How could a process which is useless or worse when put to work privately act in beneficial ways when members of a group each make use of it?

We have already seen one example of how this works, and there is nothing magical or mystical about it. Confirmation bias hides our errors from us in private moments of reflection, but when we reflect as members of a group, the very same process serves to distribute the intellectual labor of evaluating alternative solutions to the problem at hand. This is another important bit of evidence in favor of Mercier and Sperber's hypothesis, and it serves to fill out the picture of how reflection operates constructively in groups despite the fact that it does not operate constructively when individuals reflect on their beliefs without conversational partners. It would be helpful to have further evidence of how the group practice of giving and asking for reasons,[3] aided, as it is, by the reflection of individual group members, serves in the acquisition and dissemination of knowledge.

Mercier and Sperber offer a wealth of such evidence, and I can only offer a few further examples here. We have already seen, in section 4.3, that when individuals are given evidence for and against a belief they already hold, they do not deal with the evidence even-handedly. Instead, they take evidence in favor of the beliefs they hold very seriously and give it little scrutiny, while easily coming up with reasonable challenges to the evidence against their beliefs, thereby dismissing such evidence as of little value. Why should we think, then, that having a conversational partner present evidence against one's belief will make any difference? If evidence against one's beliefs is so quickly dismissed, then having a conversational partner present the evidence, rather than merely reading about it, might seem to make no real difference.

But it does make a difference, as Mercier and Sperber point out (2017, 223–36 and 267–70), and it is not difficult to see why. If I read about some evidence that one of my beliefs is false and quickly come up with some reason to challenge it, my intellectual work is done. I can go on believing as I did before I learned about this additional evidence, and I can even present my reasons for dismissing it if someone should ask. But if you believe the opposite of what I do, and you present the evidence against my belief, then when I present some quick challenge to what you say, you are unlikely to come around to my point of view if the challenge I've offered is not a good one. You will offer further reasons against my belief, reasons which serve to undermine the legitimacy of my quick dismissal of the evidence you provided, and I will need to engage with these new reasons or acknowledge that I was wrong to dismiss that evidence. When new evidence is presented on paper, it cannot respond to my challenge, and I can rest content with a quick, and perhaps shallow, response. When the new evidence is presented by a conversational partner, however, my flimsy response will not go unchallenged, and the back-and-forth which conversational partners engage in will be far more successful in testing the strength of the reasons on either side of our disagreement. Solitary assessment of reasons for belief will cut off the assessment of those reasons at an early stage, while the back-and-forth engagement of conversational partners will not only prolong that assessment, but subject the assessment of reasons to more serious scrutiny than they would receive from a lone individual whose method of assessment is so strongly biased in favor of whatever that person believed at the outset. The quality and detail of reasoning offered by groups engaged in cooperative problem-solving has regularly been found to be far better than that offered by individuals in defense of their beliefs. More than this, the quality and detail of the reasoning which emerges from group discussion regularly exceeds the performance of its best-performing members acting alone. The group does not simply take on the reasoning of its best-performing members. Instead, even that reasoning is subjected to further scrutiny, and developed in both quality and detail as a result of the back-and-forth of discussion.

Emmanuel Trouche, Petter Johansson, and Lars Hall (2016, reported in Mercier and Sperber 2017, 231–33) devised an experiment to test the ways in which scrutiny of one's own reasoning differs from scrutiny of others'. Subjects were given a number of problems to solve, and in each case they were asked to provide a solution to the problem without supporting reasons. They were then asked to explain why they provided the solution they did, and, once they did provide that bit of argumentation, they were asked to look it over and see if they wished to change their evaluation or improve upon it. Fewer than one in five of the subjects made any changes at all. Subjects were then offered solutions to those same problems, together with supporting argumentation, which, they were told, were offered by others, and they were asked to evaluate them. Among these solutions purportedly offered by others was one solution, together with supporting argumentation, that the subject him- or herself had in fact produced. Roughly half of the subjects did not recognize that they themselves had produced this particular bit of reasoning, and when these subjects were now asked whether that argument was a good one—an argument which they not only had authored themselves, but had been given an opportunity earlier to revise or improve and which they found, at that time, to be fully adequate—now about half of these subjects became more critical of the argument than they had been before. These subjects thus were quite happy with a bit of argumentation when they knew that they had produced it, but when they thought it was someone else's argument, they often thought it stood in need of improvement. When subjects did revise an argument which they now took to be inadequate, they were not just indiscriminately rejecting argumentation because they thought it was someone else's. Instead, they tended to replace weak argumentation that they themselves had offered with better reasoning. The way in which we approach reasoning and argumentation thus works best when we work in cooperative groups assessing one another's reasoning rather than trying to assess reasoning we know to be our own.

It is important to be clear about what Mercier and Sperber are claiming. They are not claiming that groups will always be able to solve problems or that they will always do better than any individual. They are not claiming that groups will never perform poorly. They are claiming, however, that when groups work cooperatively on answering some question, they will have a strong tendency to do substantially better than individuals addressing the problem alone, including the very individuals who are part of the group. Our capacity to reflect on our beliefs and the quality of our reasons works best in such cooperative social endeavors.

We can all think of situations in which groups perform very badly in addressing some issue. Most of us have, at times, been members of such groups. There are groups where some group member behaves like a bully, intimidating other members until they are all willing to go along with whatever that person thinks, rather than deal with the unpleasantness which the bully can mete out. Such groups do not do well in problem-solving. They will not, of course, outperform the lone bully, since they are cowed into submission, and, in practice, the bully is the only one who makes any contribution to the group decision. Such groups present no challenge at all to the Mercier and Sperber hypothesis, which makes a claim about groups which operate cooperatively. Groups which are dominated by a bully are manifestly not cooperative problem-solvers.

Similarly, we are all familiar with group members who are so wedded to their own opinions that no amount of argumentation against them, however cogent, will get them to change their minds. If there are enough people like this in a group, it will not perform well in problem-solving. But, once again, this is not the sort of group which Mercier and Sperber are talking about. People who do not take seriously the contributions of other group members, whatever they may be, are not engaged in a cooperative activity.

It is important to point out as well that an essential feature of the group discussions that allow for constructive engagement, if Mercier and Sperber are right, is that the initial opinions held by group members include a diversity of views. It is the disagreement among group members that exposes the various arguments offered on each side of an issue to the scrutiny needed for effective evaluation. A group of like-minded individuals will not perform well because a diversity of views is needed to make group discussion function effectively.

One might now wonder, with these points being made, whether any groups at all satisfy the conditions imagined by Mercier and Sperber. If they are only claiming that group discussion produces effective problem-solving in some exceptionally idealized situations which never actually occur, or occur only so rarely that in the vast majority of actual cases group problem-solving works very badly, then their claim is not only far less interesting than it first seemed; it is also extremely misleading. It is thus important to determine whether this is the case or not.

Mercier and Sperber are not claiming that group decision-making will not work well if a single member of a group is less than ideal, or even if every member of a group is less than ideal. We are all humans, and we are, indeed, all less than perfect. Their results, which have been extensively tested on real human subjects, are quite robust. When group members are, by and large, engaged in an enterprise which the majority is committed to, and which they regard as an attempt to figure out a solution to a problem rather than an attempt to get their own way, the Mercier and Sperber hypothesis will apply. Not all groups are like this, but many groups are. The Mercier and Sperber hypothesis is a substantive claim about such groups, and it is a substantive claim about the function of our capacity to think about and evaluate our reasons. Notice too that, insofar as their hypothesis is about what this intellectual capacity was selected for, it could not be, if they are right, that the kind of group in which this capacity produces good results does not ever exist or exists only very rarely. Such a capacity could only be selected for if good results are frequently produced in actual situations. Mercier and Sperber could, certainly, be mistaken, but they are not making a trivial claim about situations so idealized that they have no implications for the performance of actual human interactions.

5.5. A Problem Case

This is not to suggest that there are no problems at all for Mercier and Sperber's view. I think that there are some problems. In this section, I will describe a problem case which illustrates a class of difficulties for Mercier and Sperber's hypothesis.

Some problems are fairly difficult to solve, but once one sees the correct solution, it is obvious that it is correct. Consider this problem which Emmanuel Trouche, Jing Shao, and Hugo Mercier gave to experimental subjects.

> Paul is looking at Linda and Linda is looking at John. Paul is married but John is not. Is a person who is married looking at a person who is not married? (Mercier and Sperber 2017, 233)

Subjects could answer "yes," "no," or "not enough information to decide."

When subjects were given this problem individually, most thought that there was not enough information to decide. This is not, however, the right answer. To see why that is so, draw a diagram. Put Paul on the left, Linda in the middle, and John on the right. You can draw arrows pointing from Paul to Linda, to show that Paul is looking at Linda, and from Linda to John, to show that Linda is looking at John. Put an M under Paul to indicate that he is married, and an S under John to indicate that he is single. So far, we can't tell whether there is a person who is married who is looking at a person who is not married. But let's think about this for a second. We don't know Linda's marital status. Let's suppose, for a moment, that she is married. We know that Linda is looking at John, and he's single, so in that case, a person who is married is looking at a person who is not married. Of course, we conveniently supposed that Linda is married. So what if she isn't? What if Linda is single? We know that Paul is married, and he's looking at Linda. So in this case too, someone who is married is looking at a person who is not married. So no matter what—no matter whether Linda is married or single—a person who is married is looking at a person who is not married. There is enough information provided to answer our original question, and the right answer is yes.

Once one sees this bit of reasoning, the correct answer to our original question is obvious. Not many people may come up with this insight on their own, but if one is part of a group, and anyone in the group comes up with this bit of reasoning, the group is likely to recognize that this is indeed the right answer. There are a lot of problems like this, problems where the answer is somewhat difficult to figure out, but it is easy to recognize as right, or, at least, easy to recognize as right if given the proper sort of supporting reasoning. It should be no surprise at all that groups perform well on problems like this, especially large groups, since the larger the group, the more likely it is that there will be someone in the group who will figure out the right answer, and, once it is explained, most of the rest of the group will recognize that it is right.

But not all problems are like this. That, by itself, is no difficulty for Mercier and Sperber. Indeed, as Mercier and Sperber emphasize, one of the virtues of group discussion, on their view, is that it typically improves upon the reasoning of individual group members. Cases like the Linda, John, and Paul problem, while they certainly illustrate the way in which group discussion may serve to disseminate knowledge, do not illustrate some of the most important advantages of group engagement with intellectual problems.

There are intellectual problems, however, whose character is quite different from the Linda, John, and Paul problem, and which are not well served by group discussion.

When I was a child, there was a very popular television game show, *Let's Make a Deal*, hosted by Monty Hall. Contestants would be shown three curtains, and they would be asked to choose one of them; they would receive, as a prize, whatever was behind the chosen curtain. One curtain had some valuable prize behind it: a fancy car, an elaborate vacation, various kitchen appliances, or some such thing. The other two curtains had, as I remember, either nothing behind them, or some booby prize. Once the contestant chose a curtain, Monty Hall, who knew where the prize was located, would open one of the curtains which was not chosen, showing that the prize was not there. The contestant was then given a choice: stick with the curtain you've chosen, or switch to the other as yet unopened curtain. Contestants agonized over this choice as members of the audience shouted encouragement to either switch or stay with the choice they first made. This was—really!— a very popular television show years ago.

This setup led to the creation of a simple brain teaser, named after the show's host: The Monty Hall Problem. As it is typically stated, we imagine the contestant facing the choice exactly as described above. The two curtains which do not have the desirable prize behind them are sometimes colorfully imagined to each hide a goat. The contestant makes a choice—perhaps Curtain Number One—and Monty Hall shows that one of the remaining curtains—perhaps Curtain Number Two—has a goat behind it. The contestant is then allowed to stay with Curtain Number One or to switch, instead, to Curtain Number Three. What should the contestant do? Should that person stay with their initial choice or switch to the as yet unopened curtain? Or are these two options equally likely to have the prize behind them, thus providing no reason to favor one choice over the other?

Most people presented with this problem believe that there is no reason at all to favor one of these options over the other. Their reasoning, at least typically, is roughly like this. There are three curtains, and the prize is behind one of them. So the chance that the prize is behind Curtain Number One is one-third; the chance that the prize is behind Curtain Number Two is one-third; and the chance that the prize is behind Curtain Number Three is also one-third. So when the contestant makes a choice, they have a one in three chance of having chosen the prize. After they are shown that one of the other curtains has a goat behind it, they can switch to the remaining unopened curtain, but there is no more reason to believe that that curtain has the prize behind it than that the originally chosen curtain has the prize behind it. As we've said, the curtains are all equally likely to have the prize behind them. So there's no reason to switch one's choice, although there's no harm in it either. One is exactly as likely to win the prize if one stays with one's original choice as one is if one switches.

Most people find this reasoning quite compelling. It is, however, erroneous; one is actually better off if one switches when given the chance. Indeed, switching doubles one's chance of winning the prize. How could that be?

It is true that, at the outset, there is no more reason to believe that the prize is behind any one curtain rather than any other. It is equally likely to be behind each of the curtains, given what one knows. So if the contestant chooses Curtain Number One, there is a one-third chance that the prize is behind that curtain, and that means that there is a two-thirds chance that the prize is behind one of the other curtains. As a result, if one always stays with their original choice, and this game is played repeatedly, one will win the prize roughly one out of every three times. In two out of three cases, on average, the prize will be behind one of the other two curtains. Before Monty Hall reveals that the prize is not behind Curtain Number Two, if one were given the option of switching one's choice, switching, that is, to pick either Curtain Number Two or Curtain Number Three, there would be no basis for making a choice. Given what one knows, staying with one's original choice would be exactly as likely to get you the prize as switching to either Curtain Number Two or switching to Curtain Number Three.

But when Monty Hall shows you that the prize is not behind Curtain Number Two, you now know something that you didn't know before he opened that curtain, and this changes what it is reasonable to do, given what you now know. Your original choice of Curtain Number One, as we said, was no more likely to have the prize behind it than either of the other two curtains. The chance that the prize is there is just one in three. But now you know that the prize is not behind Curtain Number Two, and that means that the chance the prize is behind the remaining curtain is now two out of three. You do twice as well to switch your choice as to stick with what you originally chose.

Are you convinced?

Here's another way to think about it. When you first make a choice, your chance of winning is just one in three. Let's suppose you got lucky and you made the correct choice. Then when Monty Hall gives you the opportunity to switch, you will do worse if you do so; you'll have given up the prize and ended up with a goat. But this will happen only about one-third of the time. In two-thirds of the cases, your original choice will have been for one of the curtains with a goat behind it. And in those cases, if you switch, you will be switching from a situation in which you end up with a goat to one in which you get the prize, since you've just been shown where the other goat is. If you play this game frequently, your initial choice will have been a bad one two out of three times, and so switching to the remaining unopened curtain in those cases will get you the prize. You'll be better off switching in two-thirds of the cases. To put the point slightly differently, once you are shown which of the other curtains has a goat behind it, you are twice as likely to win if you switch as if you stay with your initial choice.

Many people are still unconvinced by this. If you are among them, I suggest you play this game with a friend repeatedly. Have the friend make up three cards, blank on one side and with a drawing on the other side of either a goat or a dollar sign. Your friend will put the cards down randomly with the blank side up, making note as she does this where the card with the dollar sign is. You then choose one of the cards, and your friend turns up one of the remaining cards with a picture of a goat on the other side. (There will always be such a card to turn over since if you've chosen the card with the dollar sign on it, both of the remaining cards will have a goat on them; and if you've chosen a card with a goat on it, there will still be one remaining card with a goat on it.) Stick with your initial choice every time, and see whether you get the prize or not. Do this repeatedly for some large number of cases, making sure that the cards are shuffled between each turn. You'll soon find that you are winning roughly one-third of the time. If this doesn't by itself convince you, then do this again and change your strategy. This time, after you make your initial choice and are then shown that one of the two remaining cards has a goat on it, switch your choice. Do this for a large number of cases. You will soon find that you are now winning roughly two-thirds of the time. You really are twice as likely to succeed if you switch your choice once you are shown where one of the goats is found.

This problem differs from the Paul, Linda, and John problem in a crucial respect. In that problem, once you are presented with the right answer and shown the reason and given the reasoning behind it, it is very easy to recognize that answer as correct. But in the Monty Hall Problem, not only is the mistaken answer intuitively compelling initially; it is also the case that the right answer is deeply counterintuitive even after being provided with an explanation for why it is right. Most people, especially before trying the experimental test of actually playing the game repeatedly, are unconvinced by any number of explanations as to why switching their choice of curtain would actually improve their chance of winning the prize.

Marilyn vos Savant discovered this the hard way. Vos Savant writes a mathematical puzzles column for *Parade* magazine, and in 1990 she presented the Monty Hall Problem to her readers. When she revealed the correct answer, together with an exceptionally clear explanation of just why one has a better chance of winning the prize if one switches one's choice when given the opportunity, she received thousands of letters, many from people with advanced training in mathematics, insisting that she was wrong. She patiently and repeatedly explained why refusing to switch actually left one less likely to win the prize, but, at least from the point of view of convincing others, she was largely unsuccessful. Her answer, though correct and carefully explained, was found unpersuasive.

Vos Savant did not have face-to-face discussions with those who disagreed with her, but she did have repeated back-and-forth written exchanges with them in subsequent columns. This is not exactly the sort of verbal interchange that Mercier and Sperber discuss, but it is, in many relevant respects, quite similar. Unlike the simple case in which one is presented with written evidence and then makes a decision, vos Savant and her readers had extensive interchanges. So it is striking that she was largely unsuccessful in convincing those who disagreed with her. At the same time, given the nature of this particular problem, it should not be surprising. The correct answer is extremely counterintuitive, and the reasoning which supports that answer is complex, especially when compared with the highly intuitive and simple reasoning which seems to support the mistaken answer.

Most importantly, the Monty Hall Problem is not unique in this respect. There are very many complex issues which seem to have a simple and highly intuitive resolution which is, in the end, incorrect. The correct solution to many of these problems is sufficiently complex as to resist easy comprehension. Convincing people to give up a belief which seems highly intuitive by way of a complicated argument which is hard to follow is an uphill battle, even for gifted teachers like vos Savant. In situations like this, group dialogue is unlikely to have the beneficial effects which Mercier and Sperber describe. Instead, in cases like the Monty Hall Problem, group discussion is likely to entrench mistaken belief.

How much of a problem is this for the Mercier and Sperber hypothesis? If the vast majority of problems we face were like the Monty Hall Problem, this would show that Mercier and Sperber are just mistaken. But even if we allow, as it clearly seems we should, that there are a great many intellectual problems which share the relevant features of the Monty Hall Problem, we are still a very long way from the suggestion that most of the problems we face share that structure. Just as we can see that the Monty Hall Problem is far from unique in having a mistaken answer which is highly intuitive and a correct answer which can only be defended by elaborate argumentation which is difficult to follow, it is also quite clear that there are very many problems we face in which correct answers may be supported by reasons that are not hard to grasp. In situations of this second sort, the evidence Mercier and Sperber offer strongly suggests that group discussion is not only helpful in arriving at the correct answer, but helpful, as well, in developing convincing supporting arguments for it.

It would be especially helpful to have a way of dividing intellectual problems into those that lend themselves to solutions, and supporting argumentation, that will strike a responsive chord in typical individuals, and those intellectual problems which do not. It would be helpful as well to have some idea of the extent to which the problems we typically face fall into each of these groupings. If most of the problems we confront are of the first sort, then the prospects for solution by way of group discussion among typical individuals are very encouraging. If not, group discussion among such individuals will often be counterproductive.

There is no question that the world is a complicated place, and many of the concerns we have can only be adequately addressed by people with relevant expert knowledge. No one would think that the problem of global warming, to take a single example, can be adequately addressed by people without highly specialized training, whether they get together in groups to discuss the issue or not. What the work of Mercier and Sperber suggests, however, is that we should not think that solving such complicated problems is best addressed even by those with specialized training working on their own. Dialogue among people without such training is generally pointless for these problems, but the benefits of group discussion which Mercier and Sperber highlight are likely to emerge in discussions among specialists trying to address such issues just as these benefits emerge in discussions of less complex problems by nonspecialists. It is a commonplace that two (or more) heads are better than one. The work of Mercier and Sperber offers us an illuminating explanation of some of the reasons why that is so.

5.6. Conclusion

We are the beneficiaries of powerful cognitive processes which allow us to gain a great deal of knowledge in private encounters with the world. Much of cognition is a matter of such private encounters. One of the striking features of human beings, however, is our capacity to engage in cooperative problem-solving with others in which we each offer our own ideas about how to approach a problem, and we critique each other's reasons in an attempt to arrive at a solution. When thinking about the extraordinary intellectual achievements of the human species, we have a tendency to think about these as the product of individual endeavors: Copernicus's development of the heliocentric model; Newton's understanding of the laws of motion; Darwin's development of an evolutionary account of the origin of species; and so on. Such a picture distorts our understanding of these achievements. As Newton famously remarked, "If I have seen further than others, it is by standing on the shoulders of giants." Newton did not arrive at his views without the benefit of the intellectual work of others, and it would be a mistake to see human intellectual achievement in general, even apart from that of geniuses like Newton, as created wholly independently, with no input from anyone else. The human species has a capacity to make use of the intellectual work of others in ways that other animals cannot. Our capacity to speak and to create written records makes it possible to stand on others' shoulders and thereby see further, even in cases where our own intellectual capacities are no better than, or inferior to, those on whose shoulders we stand. Because of this, we are the beneficiaries of the intellectual endeavors of previous generations in ways that members of other species are not. If we are to properly understand how our intellectual achievements are possible, this fact about human cognition—that it is not just a matter of lone individuals bringing their cognitive capacities to bear on the world in isolation from others—must figure prominently.

Newton's famous remark, however, understates the way in which social factors play a role in the acquisition of knowledge. You and I have books which record the accumulated products of human investigation, allowing us to see further than we ever could have if we had to address intellectual problems entirely on our own, just as Newton noted. What this leaves out, however, is the intellectual benefit that comes from the kind of cooperative problem-solving that emerges when we reason together, challenging one another's views of some matter, each revising and augmenting our reasoning in response to the reasons which others provide. It is one of the important features of Mercier and Sperber's work that it shines a spotlight on this aspect of human knowledge.

Suggestions for Further Reading

A full presentation of Mercier and Sperber's current view may be found in their book *The Enigma of Reason*, Harvard University Press, 2017. They presented an earlier version of their approach in "Why Do Humans Reason? Arguments for an Argumentative Theory," *Behavioral and Brain Sciences*, 34 (2011), 57–111. Their view changed in important ways between the presentation in 2011 and the more recent book, but the earlier article is followed by a large number of critical responses, and it is especially useful for getting a sense of alternative views of these matters, as well as potential problems in their approach. For another critical response to Mercier and Sperber, see Sinan Dogramaci, "What Is the Function of Reasoning? Mercier and Sperber's Argumentative and Justificatory Theories," *Episteme*, 17 (2020), 316–30.

The evolutionary approach which Mercier and Sperber take for granted is not uncontroversial. For a good presentation of arguments against evolutionary psychology, see David Buller, *Adapting Minds: Evolutionary Psychology and the Persistent Quest for Human Nature*, MIT Press, 2005.

And, once again, Alvin Goldman's *Knowledge in a Social World*, Oxford University Press, 1999, is essential reading for anyone who is interested in the philosophical issues surrounding the social character of knowledge.

6
Conclusion

Born to Know

6.1. How Is Knowledge Possible?

In this final chapter, I want to draw out some of the larger lessons
which have emerged from our discussion thus far. Some readers may
have wondered: where's the philosophy? We have discussed many issues in psychology about how various psychological processes operate
and what functions they perform, but I've offered this book as a contribution to epistemology, the philosophical study of knowledge. It
may seem that I've somehow lost the thread and changed the subject.
Instead of writing a book offering a contribution to the philosophical
study of knowledge, I've offered, instead, a discussion of some issues
in a completely different field of study. However interesting those issues in psychology may be, you may feel that you've been the victim of
false advertising, a kind of academic bait and switch. You were prom-
ised a book in philosophy and, instead, you were given a book that has
no philosophical content. It may be an interesting book—I very much
hope that it is—but it doesn't seem to be the book you were told you
would be getting.

I have, by design, left much of the philosophical discussion in the
previous chapters merely implicit. This is not, I acknowledge, the way
in which philosophical texts typically proceed, even texts which, like
this one, see empirical work in the sciences as deeply relevant to philosophical issues. My reason for proceeding in this way is simple. I think
it's easier to get the philosophy right if one understands the psychology
first. I believe that a good deal of philosophical theorizing has been
based on extremely intuitive views about how the mind works, views
which seemed so obvious and commonsensical that they didn't even
need to be stated. We now know that many of these commonsensical

Scientific Epistemology. Hilary Kornblith, Oxford University Press. © Oxford University Press 2021.
DOI: 10.1093/oso/9780197609552.003.0006

beliefs are false, and what that means is that our philosophical views about the nature of knowledge stand in need of serious revision. Rather than start with a discussion of those philosophical views which, as I see it, are intuitively appealing but deeply mistaken, I have chosen a different approach. I start with an account of how the mind works, and then try to build a philosophical view about the nature of knowledge on the basis of that understanding.

Now that we have reached the final chapter of the book, it is time to make the philosophy more explicit. It is time, as well, to make explicit, far more than I have done thus far, how the view offered in this book compares with other approaches and views in epistemology.

I started with the question of how knowledge is possible. This question has occupied philosophers for as long as anyone has wondered about the nature of knowledge. The bare question—how is knowledge possible?—requires some context if we are to so much as understand what is being asked. In different contexts, these very same words may be used to ask completely different questions. Someone may utter the sentence, "Why is the sky blue?" in one context as a serious question about optics, and, in another context, and with an arrogant shrug of the shoulders, as a way of dismissing as utterly pointless someone else's concern about some totally different matter. To understand what philosophers have been asking when they ask about the possibility of knowledge, we need to see that question in context.

As I argued in Chapter 1, philosophical concern about the possibility of knowledge has typically been prompted by the skeptical problem. A skeptical scenario is described, one of the nightmare scenarios described in that chapter, and an argument is offered for the conclusion that knowledge is impossible. We all believe, however, that we have a great deal of knowledge, and so we are faced with a problem. We need to be able to explain how knowledge is so much as possible given the nightmare scenarios and the skeptical argument based upon them.

I have argued that it is a mistake to try to respond to the skeptic on his own terms. That is, it is a mistake to try to convince someone who denies knowing anything at all, and denies that anyone else knows anything at all, that they are wrong about this. We can enter into rational dialogue with others, trying to convince them that they have made some mistake, only if they have some beliefs already and some forms of inference which they accept as legitimate. Without any such starting place for rational discussion, it is impossible to give someone who disagrees with you reasons for changing their mind. The skeptic offers us no such starting place, no point of entry for rational engagement. This makes it pointless to try to convince the skeptic, as has often been recognized. It also means that the fact that we cannot rationally convince the skeptic that knowledge is possible is entirely unrevealing about the nature of knowledge. Someone who stubbornly insists that the earth is flat cannot be convinced by any argument, but this, of course, shows us nothing about the shape of the earth; it merely shows us that someone who is sufficiently stubborn about anything may be beyond the reach of rational argument. Rational argumentation cannot, unfortunately, defeat all ignorance and mistake. Similarly, our inability to convince the skeptic that knowledge is possible shows us nothing about the extent of our knowledge. We should stop trying to find some angle, some entering wedge, which will allow us to defeat the skeptic on his own terms. The exercise is pointless, and the fact that it is pointless teaches us nothing about knowledge.

Many philosophers who acknowledge this point believe, despite this, that there is much to be learned by examining arguments for skepticism. The general thought, and it seems a promising one, is that by seeing where the skeptical argument goes wrong, we may learn something about the nature of knowledge. The idea here is not to try to defeat the skeptic on his own terms, but rather to explain to ourselves—individuals who are not skeptics—what mistaken assumptions the skeptical argument trades upon. There has been a great deal of careful and detailed discussion of skeptical argument throughout the history of philosophy, right up to the present day. The thought is that a careful examination of the import of skeptical scenarios and skeptical argument will reveal important philosophical lessons about knowledge.

I have not responded to this kind of work directly. Instead, I have pursued a different strategy. I have not engaged with skeptical arguments at all, but rather suggested that we may best make progress in understanding the nature of knowledge by looking at the phenomenon of knowledge itself, rather than focusing on skeptical scenarios in which knowledge is not only absent but in which it cannot be attained.

This approach motivated our examination of the psychological work on perception, inference, and so on. The value of this approach may be seen in the insight that comes from pursuing it.

In examining the phenomenon of knowledge, I began with two clear sources of knowledge: perception and inference. No one should deny that we can gain knowledge by way of perception. It is, often enough, entirely effortless. Open your eyes, and perceptual knowledge will be produced in you. We are awash in perceptual knowledge, as are young children and many other animals. When we look at how our perceptual mechanisms work, we see something very interesting about them. These mechanisms, whose presence in our species is explained by natural selection, embody certain presuppositions about typical environments. The operation of our perceptual systems trades on the fact that these presuppositions are actually true, or nearly true, of environments in which we are typically found. When these presuppositions are true, as they usually are, the perceptual systems will produce appearances which strongly incline us to form beliefs about the world around us which are themselves true. When the presuppositions are false—that is, when we are in sufficiently atypical environments—these systems will produce misleading appearances, perceptual illusions, which will incline us to form false beliefs. We can learn to work around these misleading appearances, but, leaving that aside for the moment, the perceptual systems work in such a way that they will effortlessly produce true beliefs in us in the vast majority of circumstances we find ourselves in. It is because these systems have roughly accurate presuppositions built in to them from the start, without our having to learn for ourselves about what typical environments are like, that perceptual knowledge is possible.[1]

Inferential knowledge is similarly ubiquitous. It is present not only in human adults, but in very young children and in nonhuman animals. Inference does not require self-conscious thought about what to believe on our evidence. We have seen that, like perceptual knowledge, inferential knowledge frequently occurs in us automatically, without the need for conscious supervision or attention. A good deal of philosophical work is focused on questions about how to distinguish good inferences from bad, that is, what features of an inference make it apt for producing knowledge, and what features of an inference are incompatible with the production of knowledge. Aristotle, who invented the study of logic, suggested that logic was the science of proper inference, and that very attractive idea was expanded upon with the development of probability theory to suggest that the good inferences, the kinds of inferences which could lead to knowledge, were those licensed by logic and the theory of probability. As we have seen, the key feature of valid logical inference is that if the premises of a valid argument were true, the conclusion would have to be true as well. This means that a valid argument does not owe its validity to any distinguishing feature of the world in which we live. A valid argument is valid whatever the world might be like. While probabilistic arguments with true premises do not guarantee the truth of their conclusions— they merely make them probable—they nevertheless share an important feature with valid arguments: what makes a probabilistic argument a good one does not depend on any distinguishing feature of the world we live in. A good probabilistic argument in this world is a good probabilistic argument in every imaginable world. The view on which good inference is identified with inference licensed by logic or the theory of probability thus sees good inferences as ones which would remain good in any imaginable world at all.[2]

We have seen that human inference does not work this way, and, although I have not presented the evidence for it here, inference in nonhuman animals doesn't work this way either. Just as our perceptual systems owe their reliability, and their success in producing knowledge, to certain presuppositions built into those systems, presuppositions which are at least approximately true about typical features of the environments in which we live, we have seen that a good deal of the inferences we make work in just the same way. Many of our native inferential dispositions achieve their reliability in producing true beliefs because they too make certain presuppositions about typical environments, presuppositions which are at least roughly true of the environments in which we live. These presuppositions are not true of every imaginable environment; the inferences which we are naturally inclined to make would not operate reliably, and would tend to produce false beliefs rather than true ones, in worlds very different from ours. The way in which much of our inferential knowledge of the world is achieved is thus quite unlike the inferences sanctioned by logic and probability theory. Our native inferential dispositions are adapted by natural selection to work well in producing knowledge in situations like the ones we tend to be found in, even if they wouldn't work well in every imaginable situation.

Not all of our beliefs are produced by automatic processes; we sometimes stop to reflect on what to believe, and our beliefs may, at times, be a product of such reflection. Many epistemologists have seen such self-conscious reflection as a model for how we ought to arrive at our beliefs, and a necessary condition for the production of knowledge. Even if one recognizes, as one should, that such reflection is not necessary for knowledge, it is very widely held among epistemologists that private reflection is an important vehicle for the acquisition of knowledge. Reflection is widely thought, as common sense would tell us, to serve as an important check on beliefs already held, as well as a route to the acquisition of new beliefs when automatic processes fail to provide us with answers to questions we wish to address.

We have seen that private reflection on beliefs already held does not, in fact, typically help us achieve this greater reliability. Instead, it typically makes us more confident in the beliefs we already hold whether those beliefs were formed on a reliable basis or not. In cases where we do not have any antecedently existing belief, and we reflect in order to resolve some issue of concern to us, we are left with the vivid impression that our reflection has made our reasons fully transparent to us: we seem to have direct knowledge of the reasons on the basis of which we formed our newly acquired belief. Here too we have seen that private reflection does not work the way it seems, and that many of our reasons for belief are not revealed to us when we reflect. More than this, many of the reasons we seem to have for arriving at our beliefs, as reflection would have it, are not among the reasons which actually influence us. Private reflection is deeply misleading about many matters that are absolutely central to understanding the nature of reflective knowledge.

If private reflection is so misleading on such matters, what is it good for? A promising suggestion here comes from the work of Mercier and Sperber: reflection is not adapted for private use; it is, instead, adapted to allow us to work cooperatively with others in social situations. When we engage with others in trying to solve a problem, our capacity to reflect allows us to offer reasons to one another, and to critically assess the reasons which others offer us. As Mercier and Sperber argue, the capacity to reflect is conducive to the production of knowledge when it is employed in such cooperative problem-solving. The capacity to reflect works well in these situations, but it often functions quite badly when it is taken out of such situations and applied to situations it was not adapted for.

There is a common theme here which explains how it is that knowledge is possible. Knowledge is possible because we are natively endowed by natural selection with processes which are adapted to certain features of the environments in which we are typically found, processes which tend to lead to true belief when they operate within those environments. These processes would not produce knowledge in every imaginable environment. Creatures with cognitive equipment like ours but placed in a world very different from our own might well end up with little or no knowledge. We don't live in those worlds though, and these features of our native endowment make knowledge possible for us, as well as other naturally evolved creatures. It is by virtue of the fit between our native cognitive processes and certain widespread features of the environments in which we live that knowledge is possible.

Aristotle remarked that humans by nature desire to know. The view I have defended here says nothing about whether we desire to know, but it does amount to the view that we are, in an important sense, born to know. Natural selection has built us, and many other animals as well, so as to gain knowledge of the world around us. We could not survive without such knowledge. Our ability to function at all, to serve our most basic biological needs, depends on our ability to pick up information about our surroundings so as to act on those needs in ways that allow us to satisfy them. Many philosophers have argued that knowledge is a kind of achievement, the product of an effortful activity. On the view presented here, we should see knowledge, instead, as a natural outgrowth of our biological makeup.[3] While that makeup certainly does not ensure our infallibility or make effortful conduct in the service of knowledge irrelevant, it does ensure that without any effort at all, knowledge will be produced within us. Knowledge is not only possible for us; it is inevitable.

You may suspect that my answer to the question of how knowledge is possible does not address the question which many other philosophers were asking when they spoke or wrote those very same words. My approach certainly does not allow us to provide an argument which will convince the skeptic that we do have knowledge; but nothing can do that. And, as I have emphasized, my approach does not provide us with a reason to think that there is a way of arriving at one's beliefs which would be successful in producing true beliefs, or even successful in producing beliefs which, in some significant sense, are likely to be true, no matter what the world might be like. I have argued that this too is something which cannot be done. Our successes in gaining knowledge are not achieved in this way. Some have sought an explanation of how knowledge is possible which anyone who has knowledge would be in a position to provide for themselves. Such an explanation would reassure anyone who is at all concerned about whether knowledge is actually possible, or someone who simply wishes to understand, given that knowledge is possible, what makes it so. But there is no guarantee that knowers will have the relevant information to provide themselves with such reassurance. Indeed, the vast majority of knowers will not be in such a position. Nonhuman animals and very young children do not even have the relevant concepts available to them to explain the possibility of knowledge, but they are knowers nonetheless. And the vast majority of human beings are not aware of the relevant facts about how perception, inference, and so on work, or the evolutionary basis of our knowledge-gathering capacities. Without a knowledge of these matters, they are in no position to explain to themselves what makes knowledge possible. As a result, my answer to the question of how knowledge is possible does not provide many of the things which philosophers have sought in asking that question.

I don't believe that this shows that I have somehow changed the subject or that I am answering a different question. What it does show is that others who have tried to answer this question have assumed that the answer would be of a certain form, that it would satisfy certain conditions, or that it would be able to do certain sorts of explanatory work in addition to explaining how knowledge is possible. My answer doesn't provide the kind of answer which these philosophers hoped to find, but that is because, if I am right, the right answer to our question about the possibility of knowledge proves to be quite different from what many other philosophers thought it might be. My approach doesn't answer a different question than these philosophers were asking; it simply gives a very different answer than the ones these philosophers expected to find.

6.2. What Is Knowledge?

Many philosophers begin their study of the nature of knowledge by providing a definition of 'knowledge.' Without defining our terms, one might think, we don't even know what our inquiry is about. On such a view, a proper inquiry starts off with a definition of key terms, and we can then meaningfully ask questions using those terms to advance that inquiry. We begin our study of epistemology, on this view, by defining what we mean by the word 'knowledge,' and we may then ask various questions about knowledge, such as how knowledge so defined is possible.

I have not proceeded in this way, and it may seem that there is something odd, and something defective, about the way in which this entire study has proceeded. How can we even ask questions about knowledge if we have not begun by saying what it is we mean by that word? Isn't there something horribly shoddy about a philosophical work on knowledge which doesn't begin by defining its terms, and, instead, marches merrily along through five chapters talking about all sorts of features of something, knowledge, when there is not even the slightest attempt, at the very beginning of the study, to say what one is talking about?[4]

I do not think that the right way to begin a study of knowledge is by providing a definition of that term. My view, instead, is that one is only in a position to say what knowledge is after one's investigation has made considerable progress. An account of what knowledge is can only be provided after careful investigation of the phenomenon. It is the product of such an investigation, and it cannot precede it.

It is worth noting that, in the sciences, a proper definition of terms is the product of extensive investigation rather than something which occurs before the investigation begins. We noted this in section 2.1. Consider the investigation of the phenomenon of heat. Long before it was understood that heat is a form of kinetic energy, it was certainly well known that a summer's day in North Africa is hotter than a winter's day in the north of Sweden. Simple facts of this sort together with even a very rough-and-ready recognitional capacity for the phenomenon of heat allowed investigation of that phenomenon to begin. Early on, it was thought that heat might be a fluid contained in physical objects: hotter objects had more of the fluid, while colder objects had less of it. As investigation proceeded, it became clear that this was not the case, and our understanding of the phenomenon increased. A definition of heat in terms of kinetic energy was a late arrival on the scene. It could not possibly have been given in advance of the investigation of the phenomenon.

I have proceeded in my investigation of the nature of knowledge in much this way. I have not tried to define my terms before investigating the phenomenon. Instead, I have taken for granted that such a definition could only be given after making substantial progress on that investigation. This does not mean that the investigation of that phenomenon was somehow ill-founded or shoddy, any more than the investigation of heat was ill-founded or shoddy because physicists did not produce a definition of 'heat' before they so much as started to look at the phenomenon.

It is important to note that although the subject matter of an investigation needs to be pinned down somehow if it is even to begin, the pinning down need not be done by way of a definition of key terms. One can point out examples of the phenomenon one wishes to study, as physicists did with heat, and one can indicate various features of the phenomenon which cry out for explanation. Notice that we began, in Chapter 2, by giving a variety of clear examples of the phenomenon of knowledge, and this was sufficient to prompt more detailed investigation of various features of that phenomenon. This is what led us to look at psychological studies of perception and inference, as well as reflection and the social practice of giving and asking for reasons. These studies revealed interesting facts about the phenomenon of knowledge which could inform our account of what knowledge is, features which we could not have taken into account if we tried to define the term before looking at those psychological studies.

It is time, however, to propose an account of what knowledge is, even while acknowledging that any such account is necessarily provisional. An account of knowledge, like an account of heat, may need to be modified in light of new information. New discoveries are constantly being made, and they may force us to revise our account in important ways. I do think, however, that enough is known at this point to offer an informative account of the nature of knowledge, one we could not have offered prior to examining the phenomenon in the way that we have in the chapters of this book. I believe that account is actually implicit in much that has already been said in those chapters, but it is time to make it explicit.

I follow Alvin Goldman (1986, 2012) in holding that knowledge is just true belief which is the product of a reliable process. What makes a belief a case of knowledge is nothing more nor less than that it is true, and that it is produced or retained by a reliable process. I will have a good deal to say about what this view comes to, both in this section of this chapter and the next section, but I want to begin by highlighting a difference between Goldman's view and my own. I agree with Goldman about what knowledge is, but we disagree about how we may come to know that this is the right account of the nature of knowledge.

Goldman favors a traditional philosophical methodology in which an account of what knowledge is, or indeed, a philosophical account of anything at all, is tested against our intuitions about a variety of hypothetical cases, including highly imaginative cases involving situations very different from anything that might actually occur in the natural world (Goldman and Pust, 1998; Goldman 1986, 2005, 2007). Suppose a philosopher proposes that knowledge is nothing more than true belief; any true belief, on such a view, counts as knowledge. One might show that such a view is mistaken by considering a case in which someone comes to believe something which is in fact true—perhaps that their favorite presidential candidate will win in the coming election—simply because it makes them feel good. If we think about this case for even a moment, most everyone will be struck by the very strong impression—what philosophers call an intuition about this imaginary case—that such a belief about the outcome of an election, even should it turn out to be true, would not count as knowledge. The methodology which Goldman favors treats such intuitions as evidence for or against various philosophical theories. Intuitions, on this view, play the role in philosophical theories that observations play in the sciences. Hypotheses in the sciences are tested against experimental observations; hypotheses in philosophy, on Goldman's view, are tested against our intuitions about hypothetical cases, including hypothetical cases involving situations utterly different from anything we might actually encounter.

Goldman defends his view that knowledge is reliably produced true belief by way of this methodology. My own view about proper method in philosophy is quite different.[5] As I have been arguing, proper method in philosophy is not so different from proper method in the sciences. We find some phenomenon which catches our attention and which we seek to understand better. We begin by looking at what seem to be clear cases of that phenomenon and we try to figure out, by examining the phenomenon itself, what it is that these cases have in common. We don't consider wild hypothetical cases which are utterly unlike anything that might arise in nature; rather, we look at the kinds of cases which actually occur and subject them to scrutiny. This is how chemists and physicists were able to come to an understanding of what water is, and what gold is, and what heat is. What I have been suggesting is that we may proceed in just the same way in philosophy, and that is what I have been doing here with the case of knowledge. There is a robust phenomenon of human knowledge, both in adults and in children; there is a robust phenomenon of knowledge in large swaths of the animal world. I have suggested that we examine various instances of this phenomenon, and that is why we have looked at perceptual knowledge, inferential knowledge, the process of reflection, and the ways in which the exchange of reasons with others may result in knowledge. We examined all of these cases to see what different instances of knowledge have in common.

The analogy with scientific methodology may strike you as odd. In the scientific cases, one might think, the subject matter under investigation is some natural phenomenon or some natural kind of stuff, such as water or gold. But knowledge, some will say, is not like that. Some philosophers have suggested, for example, that knowledge is importantly different form these various natural phenomena or natural kinds in that it has a normative dimension. In order for a belief to count as knowledge, these philosophers suggest, our reasons must meet certain standards. We must have good reasons for belief, if our beliefs are to count as knowledge; our reasons must be adequate; they must be good enough. But these standards, what counts as adequate, or good, or good enough, are not determined by something in the world that we might discover in the way that we might discover what makes something water, or gold, or heat; standards for knowledge are something that we impose upon the world rather than something that we discover out there independent of us.[6] And what this means, if this line of thought is correct, is that an important feature of knowledge, its normative or evaluative dimension, is not to be found out in the world in the way that the objects of scientific investigation are found; a crucial feature of knowledge is to be found in us. It is for this reason, one might think, that methodology in at least those parts of philosophy which deal with normative issues, issues about how things ought to be and not just issues about how things are, must be different from proper methodology in the sciences. Because knowledge has a normative dimension, because it requires the possession of adequate, or good, or good enough reasons, it is not appropriate to pursue it by the same sorts of methods which we apply in the sciences.

My view, however, is that knowledge is, in fact, a natural phenomenon susceptible to the very kinds of investigative procedures we pursue in the sciences. Knowledge is an object of study in the cognitive sciences, including psychology, neuroscience, linguistics, cognitive anthropology, cognitive ethology, and parts of sociology. There is a large issue here, but let me try to make the case for this approach briefly.

As I have been emphasizing throughout this book, knowledge is found not only in human beings but in many nonhuman animals. In the scientific study of the cognitive capacities of different species, just as in the scientific study of the biology of different species, an evolutionary approach plays an important role. Texts devoted to cognitive ethology, the study of cognition in nonhuman animals, are typically organized in ways that give these evolutionary considerations a prominent role. The obvious importance of behaviors connected to feeding, fighting, flight, and sexual reproduction in many species are all illuminated by approaching them in this way. The human species as well, of course, is a product of evolution, and there is just as much reason to think that an understanding of our cognitive capacities may be illuminated by viewing them in this light.

Interestingly, cognitive ethologists regularly talk about knowledge in many nonhuman animals. One might think that this is just casual, sloppy talk which plays no essential role in their theorizing, in much the same way that astronomers may certainly talk about the sunrise without literally believing that the sun rises in the morning, rather than that it merely appears to rise as a result of the rotation of the earth. Talk of knowledge in the cognitive ethology literature is not like that, however. Rather, it is deeply connected with an evolutionary understanding of the cognitive capacities of various creatures and the ways in which these play a role in influencing animal behavior. Consider the following passage from Louis Herman and Palmer Morrel-Samuels in their work on dolphins.

> Receptive competencies support knowledge acquisition, the basic building block of an intelligent system. In turn, knowledge and knowledge-acquiring abilities contribute vitally to the success of the individual in its natural world, especially if that world is socially and ecologically complex, as is the case for the bottle-nosed dolphin. . . . Among the basic knowledge requisites for the adult dolphin are the geographic characteristics and physiographic characteristics of its home range; the relationships among these physical features and seasonal migratory pathways; the biota present in the environment and their relevance as prey, predator, or neutral target; the identification and integration of information received by its various senses, including between an ensonified target and its visual representation; strategies for foraging and prey capture, both individually and in social units; the affiliative and hierarchical relationships among members of its herd; identification of individual herd members by their unique vocalization and appearance; and the interpretation of particular behaviors of herd members. . . . This is undoubtedly an incomplete listing and is in part hypothetical, but is illustrative of the breadth and diversity of the knowledge base necessary to support the daily life of the individual dolphin. Similar analyses could be made of knowledge requirements of apes or of other animal species, but the

underlying message is the same: extensive knowledge of the world may be required for effective functioning in that world and much of the requisite knowledge is gained through the exercise of receptive skills. (1990, 283–84)

Notice what is being claimed here. Animals inhabit complex environments, whether those environments be the oceans in which dolphins live, the jungles and rainforests inhabited by various apes, the vast plains which are the home of buffalo, giraffes, zebras, and so on, the air, land, and water inhabited by various bird species, or the underground homes of assorted burrowing animals. Surviving in these environments requires a sensitivity to the constantly varying conditions the environments present. To put it slightly differently, these environments make informational demands on the animals living in them. Because living creatures are a product of natural selection, they are adapted to the environments in which they live. If we want to explain the cognitive capacities we find in various creatures, we must see them as selected for their capacity to provide those creatures with the information needed to inhabit those environments successfully, and what that means is that these cognitive capacities allow the creatures to act on accurate information about their environments so as to satisfy their biological needs.

If we want to explain the behavior of an individual animal on some single occasion—perhaps the attempt by a hawk to swoop down from a treetop and capture a passing chipmunk for its dinner—we need only appeal to what that hawk believes about its environment, together with the fact that the hawk wants something to eat.[7] While the hawk's beliefs about its surround will typically be true, the hawk's behavior will be just the same whether its beliefs are true or false. So long as the hawk believes there is a chipmunk in the field when it is hungry, it will swoop down from its perch in the trees toward the place it believes a chipmunk to be. But if instead of trying to explain the behavior of this one hawk on this one occasion we wish to explain why it is that hawks in general have the cognitive capacity to recognize chipmunks at a distance, we will need to appeal to the way in which natural selection has adapted the species so as to successfully operate within its environment. But now what we are explaining is not just the fact that a single hawk has acted on its belief, whether true or false. We are explaining the fact that hawks are able to successfully engage with their environment in ways that allow them to survive, and what this requires is that they have a reliable capacity to form true beliefs about that environment. The reason cognitive ethologists like Herman and Morrel-Samuels need to talk about the acquisition of knowledge and not just the acquisition of belief without regard to whether those beliefs are true or false is that they are not just trying to explain why one animal on one occasion moves the way it does; instead, they are trying to explain why a species is able to survive in its environment, and this requires appealing to the capacity of the species to gain knowledge.

The appeal to knowledge here is thus motivated in the very same way as the appeal to any other scientific category: knowledge plays an explanatory role[8] in the theories required to make sense of the behavior of animal species. Just as physicists appeal to the properties of various subatomic particles to explain the results of certain experiments, and this gives us reason to believe that there are such particles having those properties, and just as chemists appeal to various features of the chemical bond in order to explain why certain substances combine with others in particular ways, and this gives us reason to believe that there are such chemical bonds operating so as to produce various chemical reactions, cognitive ethologists appeal to a certain sort of psychological state—knowledge—in order to explain how animal species may survive in their complex environments. If we want to understand the nature of subatomic particles, we need to look to what those theories which explain the results of various experiments say about those particles; and if we want to understand what a chemical bond is, we need to look to what the various successful theories about the chemical bond say it is. Similarly, if we want to understand what knowledge is, we need to look to those theories offered by cognitive ethologists, the theories which required an appeal to a distinctive sort of psychological state, to see what that psychological state amounts to. What we see in those theories, given the explanatory work that the notion of knowledge is called upon to do, is that knowledge is reliably produced true belief.

It may seem odd to look to scientific theories to explain what knowledge is when the very theories being appealed to, drawing on the theory of evolution by way of natural selection, are of relatively recent vintage, given that we have had a concept of knowledge for far longer than we have had a theory of evolution based on natural selection. If what makes something knowledge is a matter of reliable cognitive capacities selected for answering to the informational demands placed on a species by its environment, how could we have talked and thought about knowledge for so long before we knew anything at all about natural selection?

In fact, this kind of situation is not at all unusual. Human beings thought and talked about water long before they understood that what it is to be water is just to have a certain chemical structure, namely, to be H_2O. We were able to recognize water on the basis of certain superficial properties—it is clear, colorless, odorless, and so on; it falls from the sky as rain, and fills our lakes, and rivers, and oceans—prior to understanding what it is that makes water the kind of stuff it is. We had a recognitional capacity for water long before we had a theoretical understanding of what makes water water. This is not a rare phenomenon at all. Much of what we talk about and think about in everyday life is like this. We do not find it surprising that human beings should have been able to talk and think about water prior to understanding its true nature—what it is that, at bottom, samples of water all have in common, and what makes water the kind of substance it is. What I am suggesting here is that knowledge is just like that. Our understanding of the nature of knowledge is of recent vintage, even if we were able to successfully talk about knowledge and think about knowledge prior to having that understanding.

It may seem odd, as well, to see philosophical understanding of the nature of knowledge as dependent on scientific understanding in the way suggested here. Didn't Plato and Aristotle, as well as a host of other philosophers, present philosophical accounts of knowledge prior to Darwin's presentation of the theory of natural selection in 1859? If a proper account of knowledge requires a certain scientific understanding, and that understanding wasn't available until the middle of the nineteenth century, what should we think all these pre-Darwinian philosophers were doing?

There is nothing at all odd, however, about seeking an understanding of some phenomenon prior to knowing just what will be required to gain that understanding. How, indeed, could it be otherwise? Some phenomenon strikes our attention and we seek to understand it better. At first, we may have very little sense of just what it takes to explain that phenomenon. It may turn out to be quite easy to explain, or it may be that as we investigate it further, it turns out to be far more complex than we ever imagined. A proper scientific understanding of many phenomena has taken centuries or even millennia; in other cases, we have yet to achieve an understanding of phenomena which have caught our attention. Before we have a rich understanding of some phenomenon, however, we begin with common-sense judgment or fairly primitive theorizing, and we do what we can with that, together with further thought and examination to prod our theorizing along. Philosophers who have thought about the nature of knowledge over the years have begun with common sense, with obvious truths about some of what we know and some facts about how we know things, as well as what psychology has had to offer in further explaining how it is that we are able to achieve the knowledge we have. As our understanding of human psychology and the psychology of other animals has improved—and it has grown by leaps and bounds over the past half-century—the opportunity for deepening our understanding of the nature of knowledge has grown as well. This trajectory, from common-sense judgment about a phenomenon to richer and more scientifically informed theorizing, is seen in many areas of philosophy. The philosophical study of language, which has been with us since ancient times, has been dramatically enriched by the scientific study of language and language acquisition. Much current theorizing in the philosophy of language would have been impossible without these scientific advancements. Philosophical thought about the nature of the human mind has been similarly enriched by scientific advancements in psychology, neuroscience, computer science, and linguistics. Moral and political thought has been influenced in important ways by the scientific study of moral development and empirical studies of well-being. Philosophy has always been influenced by scientific developments, not only because such developments often raise philosophical issues of their own, but because they frequently allow us to understand phenomena of philosophical interest in a deeper and more illuminating way. Our ability to illuminate the theory of knowledge by recent advancements in science is thus of a piece with what we see throughout the history of philosophy.

Let me return, if only briefly, to the thought that knowledge has a normative dimension—to know something is to meet certain standards—and this normative dimension makes knowledge a different sort of thing than the kinds of categories which can be studied by the sciences. The thought is that scientific kinds, like water and gold and so on, are natural kinds; we discover what makes them the kinds of things they are. Kinds which involve normative standards, on the other hand, are not made by nature, but by us. We impose those standards on the world, rather than discover them existing in the world independent of us. This line of thought has had a powerful influence on philosophical thinking not only about epistemology, but about many other philosophical subjects as well, and it is a line of thought which we ought to resist. There are two different strands to this idea. There is, first, an idea about the nature of categories which have a normative dimension, that what makes them the kind of thing they are is not some feature of the world apart from how we think about it, but something that our thoughts or preferences impose on the natural world. And second, there is the idea that, because categories with a normative dimension are not part of the natural world apart from our thoughts or preferences, they are not susceptible to scientific investigation. We should resist both of these thoughts.

We have seen that there is a flourishing area of scientific investigation which seeks to explain how it is that animals—including human beings—are able to survive in the complex environments in which they are found. This work draws very centrally on evolutionary ideas, and it applies evolutionary thinking to the study of psychological processes just as much as to the study of the bodies and organs of animals. The fact that this scientific research program has proven fruitful, and that it has proven illuminating about the nature of knowledge, should make us suspect that there must be something wrong with the idea that the normative dimension of knowledge makes it immune to scientific investigation, and that with the idea that the category of knowledge is not a natural category. Knowledge cannot be immune to scientific investigation since we have a successful body of scientific work which is carrying out just such an investigation! It cannot be that knowledge is not a natural category, a feature of the world apart from our thoughts about the world or our preferences, since it is our theorizing about the natural world that has led us to view knowledge as just such a natural category!

What should we make then of the argument that knowledge involves meeting certain standards, standards imposed by us, rather than being something discovered in the world? The natural world is shot through with standards set by the world itself rather than by our thinking about the world or our preferences for how it might be. Knowledge is a matter of true belief reliably produced. How reliable must a belief-producing process be in order for the resulting belief, should it be true, to count as knowledge? One might think that this standard is up to us. Some people might set a very high standard, while others set a lower standard for knowledge. Some cultures might set one standard, while others set another. But the world, on this way of thinking, can't set a standard.

The world does, however, set a standard for knowledge. Remember that we were led to see knowledge as a scientific category because we need to appeal to this category in order to explain how creatures are able survive in a complex world. The environment a species inhabits creates certain informational demands; without an ability to pick up information about that environment, a species cannot endure. How reliable, then, must an animal's psychological processes be in picking up information about the world if those processes are to count as capable of producing knowledge? They must be reliable enough to allow the species to survive in that environment. This is not a standard which we have somehow imposed on the world because we care to have beliefs which are at least this reliably produced. It is a standard set by nature. Creatures with processes that meet such a standard will survive; those which don't, in W. V. Quine's memorable words, "have a pathetic but praiseworthy tendency to die before reproducing their kind" (1969, 126).

We thus see that although the category of knowledge does involve meeting a certain standard—in particular, it involves having beliefs which are produced by processes which are sufficiently reliable—that standard—the level of reliability required—is not imposed by us; it is imposed by nature. Knowledge is thus properly viewed as a natural category, fully amenable to scientific investigation.

6.3. Conclusion

There are very many questions in epistemology which we have not touched on here. Even an introductory discussion of all of the issues which epistemology addresses cannot be adequately provided within the covers of a single book. What I have tried to do here is present some central questions about knowledge and develop an approach to them which proves illuminating. I began the book by asking what a theory of knowledge is and why we need one. I hope to have shown that the phenomenon of knowledge is ripe for theoretical exploration, and that a theory of knowledge is well worth having.

Suggestions for Further Reading

I have developed the approach to knowledge outlined here in further detail in *Knowledge and Its Place in Nature*, Oxford University Press, 2002. Ruth Millikan's *Language, Thought, and Other Biological Categories*, MIT Press, 1984, and her *White Queen Psychology and Other Essays for Alice*, MIT Press, 1993, were pioneering works developing an approach very much in the spirit of the present work.

6.6. Conclusion

Notes

Preface

1. BonJour 2002, 2010; Feldman 2003; Fumerton 2006; Goldman and McGrath 2015; Nagel 2014; Sosa 2017; Steup 1996; Williams 2001; Zagzebski 2009.

Chapter 1

1. For some indication of the role which skeptical argument has played in epistemology over the course of the last hundred years, see, in addition to the works cited in the "Suggestions for Further Reading" at the end of this chapter, Austin 1962; Ayer 1940; Moore, "Proof of an External World," reprinted in his 1959; Russell 1912; Williams 1996; Wittgenstein 1969.
2. One might think that the skeptic does have at least one belief, namely that no one knows anything at all. We need not, however, suppose that the skeptic believes this. Rather, the skeptic may be regarded as someone who offers us—people who do, indeed, have lots of beliefs—considerations which we find problematic, and which we see as giving us compelling reason to stop believing as we do. There are complicated issues here. For an important discussion of them, see Burnyeat 1983.
3. The following passage, from Korsgaard 1996, is representative:

> I perceive, and I find myself with a powerful impulse to believe. But I back up and bring that impulse into view and then I have a certain distance. Now the impulse doesn't dominate me and now I have a problem. Shall I believe? Is this perception really a *reason* to believe? . . . The reflective mind cannot settle for perception. . . . It needs a *reason*. Otherwise, at least as long as it reflects, it cannot commit itself or go forward.
>
> If the problem springs from reflection then the solution must do so as well. (93)

Chapter 2

1. The points made in this paragraph are connected in important ways to certain issues in the philosophy of language about how it is that terms like 'water' refer to kinds of substances, and, more generally, about how words refer to objects, kinds, and properties. In short, the issues addressed here bear on questions about how it is that we may use language to talk about the world. Pioneering work on this issue was done by Hilary Putnam in a series of papers, including "Is Semantics Possible?," "Explanation and Reference," and "The Meaning of Meaning," collected in his 1975, and by Saul Kripke in his 1980.

2. Matters of epistemic agency have played a very large role in the work of Ernest Sosa, and they are the central focus of his 2015. A very different set of issues about the role of agency in knowledge is the subject of Richard Moran's 2001.

3. The distinction between reflective and unreflective knowledge has been central to Ernest Sosa's work for quite some time. See his 1991, 2007, 2015. A related distinction has occupied psychologists under the heading of the dual-systems, or dual-processes, view. See Evans 2010; Evans and Frankish 2009; and Kahneman 2011.

4. Pioneering work on the import of such social features may be found in C. A. J. Coady's 1992. Coady's book led to an explosion of work on social factors in knowledge. Particularly valuable here is Goldman 1999a.

Chapter 3

1. Those who are fortunate enough to have two functioning eyes are in a position to pick up information about the world around them in ways that someone with a single eye cannot. For those with two eyes, the images cast on each of that person's retinas are not mere duplicates, alike in every last detail. Because the two eyes are focused on objects from slightly different angles, the images cast on the two retinas differ, even if only slightly, in important ways. Indeed, the differences are sufficient to allow a good deal of information about the shape of the three-dimensional source of the retinal images to be extracted from them. But this is not the only source of information which the visual system uses to determine the shape of the three-dimensional objects in the world. There are, indeed, a very large number

of such sources, thereby allowing those with only a single eye, in spite of the limitations that imposes, to gain very substantial information about the three-dimensional world automatically and without any need for conscious attention to the task of extracting that information. For details, see Howard 2012.

2. The question of how psychological work on the kinds of inferences that people tend to make bears on the philosophical issue about the kinds of inferences that people ought to make, with special reference to the work of Tversky and Kahneman, is well discussed by Edward Stein in his 1996.

3. This point is made in Goodman 1973, which provides a pathbreaking discussion of these issues.

4. Careful readers will note that, throughout this section, I have moved quite quickly from talk of reliable inference to talk of good inference. This is, of course, a nontrivial move. I discuss this explicitly below in section 6.2. The relationship between reliability and justified belief, as well as the relationship between reliability and knowledge, is the subject of Alvin Goldman's important work in his 1986 and 2012.

Chapter 4

1. There are interesting, and extremely complex, issues at stake here. They are a focus of attention in Marušić 2015 and Moran 2001, as well as in the highly influential and formative work of Anscombe 1957.

2. As will become clear over the course of this chapter, I do not myself believe that the first-person perspective deserves the sort of deference which philosophical tradition has afforded it. Indeed, it is here that one of the great advantages of allowing our investigation of epistemological issues to be informed by scientific results may be seen. Without the benefits of the kind of psychological research described in this chapter, our epistemological views would be shaped by the illusions and distortions which the first-person perspective creates.

3. An extremely useful discussion of the role of the notion of guidance in shaping epistemological theories may be found in Goldman 1999b.

4. The example of reasoning in the game of chess played an important role in the early history of artificial intelligence. For a useful discussion of that history, as well as background on the ways in which our reasons for belief and choice cannot all be present to consciousness, see Haugeland 1985.

Chapter 5

1. Much of the credit for this newfound focus in the literature of analytic epistemology is due to the work of Alvin Goldman; see especially Goldman 1999 and Goldman and Whitcomb 2011. Within social epistemology, there has been a tremendous focus recently on issues involving the ways in which knowledge may be conveyed by way of testimony, and the complications this presents when we discover that someone else, no less reliable, in general, than we are ourselves, has a belief contrary to our own. On the important role of testimony in conveying knowledge, see especially Coady 1992 and Lackey 2008. On the epistemological significance of disagreement, see Lackey and Sosa 2006 and Christensen and Lackey 2013. I have chosen to focus my attention in this chapter elsewhere because I believe that the phenomena discussed here involve a far deeper social engagement than typically arises in the important cases of testimony and the disagreements which it may make manifest.

2. We are not, of course, the only species that engage in joint cooperative activity, and none of the remarks here are meant to suggest that such activity is impossible without the capacity to reflect on one's reasons and to articulate them. Rather, the point here is simply that such a capacity offers a substantial advantage in the coordination of plans. For a useful discussion of this point, see Tomasello 2014 and 2019.

3. The fundamental epistemological importance of the social practice of giving and asking for reasons has been emphasized by Sellars 1963, and developed in great detail by Brandom 1994. The importance which Sellars and Brandom attach to this practice, however, is of a very different kind than that suggested by Mercier and Sperber, and neither Sellars nor Brandom shows any interest in an empirical investigation of that practice or in examining what the practice actually achieves.

Chapter 6

1. This point bears on the issue dividing internalists and externalists in epistemology. For a quick overview of that issue, see the introduction to Kornblith 2001.

2. This claim about the robustness of good inference—that a good inference in any one world is allegedly a good inference in all imaginable worlds—is connected to an alleged feature of how we can come to know that inferences

which are good are, in fact, good, namely, that such knowledge is not dependent on knowledge of any features peculiar to the world we happen to inhabit. This raises the very large topic of a priori justification and knowledge. For an excellent discussion of this issue, see Casullo 2003, and the essays collected in Moser 1987. This issue about how we may know good inferences to be good is also connected to the viability of foundationalist approaches to epistemological questions. On this issue, see any of the texts cited in note 1 to the preface.

3. Here I am deeply indebted to Millikan 1984 and 1993; Godfrey-Smith 1996; and Sterelny 2003.

4. Jackson 1998 gives a very clear articulation and defense of such a view.

5. I have critically discussed Goldman's view on methodology in Kornblith 2007.

6. This way of thinking about things is vividly presented in Hume 1888 (originally published in 1739). Goldman suggests a similar motivation in his 2005.

7. Some will find talk of beliefs in hawks somewhat jarring. Should we really think that hawks have beliefs? At a minimum, we must allow that hawks, and many other nonhuman animals, mentally represent various aspects of the world. We cannot explain their behavior, and their many successes in satisfying their needs, without supposing that they have such mental representations. There is reason, however, for thinking that humans are not the only animals who have beliefs, even if exactly where we should draw the line between those creatures who have genuine beliefs and those who have mental representations of features of the world but do not have beliefs is not entirely clear. For a very useful discussion of this issue, see Allen and Beckoff 1997. I discuss this issue in chapter 2 of my 2002.

8. Williamson 2000 also emphasizes the explanatory role which appeals to knowledge play, although he focuses on this role in everyday explanatory practices rather than connecting it with any theoretical role in the sciences.

References

Abbott, Edwin. 1992. *Flatland: A Romance of Many Dimensions*, Dover. First published in 1884.

Alcock, John. 2013. *Animal Behavior: An Evolutionary Approach*, 10th edition, Sinauer Associates.

Allen, Colin, and Bekoff, Marc. 1997. *Species of Mind: The Philosophy and Biology of Cognitive Ethology*, MIT Press.

Anscombe, G. E. M. 1957. *Intention*, Blackwell.

Austin, J. L. 1962. *Sense and Sensibilia*, Oxford University Press.

Ayer, A. J. 1940. *The Foundations of Empirical Knowledge*, Macmillan.

Bertrand, Marianne, and Mullainathan, Sendhil. 2004. "Are Emily and Greg More Employable Than Lakisha and Jamal? A Field Experiment on Labor Market Discrimination," *American Economic Review*, 94, 991–1013.

BonJour, Laurence. 1985. *The Structure of Empirical Knowledge*, Harvard University Press.

BonJour, Laurence. 2010. *Epistemology: Classic Problems and Contemporary Solutions*, Rowman and Littlefield. 1st edition 2002.

Brandom, Robert. 1994. *Making It Explicit: Reasoning, Representing, and Discursive Commitment*, Harvard University Press.

Brownstein, Michael, and Saul, Jennifer, eds. 2016. *Implicit Bias and Philosophy*, 2 vols., Oxford University Press.

Buller, David. 2005. *Adapting Minds: Evolutionary Psychology and the Persistent Quest for Human Nature*, MIT Press.

Burnyeat, M. F. 1983. "Can the Skeptic Live His Skepticism?," in M. F. Burnyeat, ed., *The Skeptical Tradition*, University of California Press, 117–48.

Casullo, Albert. 2003. *A Priori Justification*, Oxford University Press.

Christensen, David, and Lackey, Jennifer, eds. 2013. *The Epistemology of Disagreement: New Essays*, Oxford University Press.

Coady, C. A. J. 1992. *Testimony: A Philosophical Study*, Oxford University Press.

Descartes, René. 1993. *Meditations on First Philosophy*, 3rd edition, Hackett Publishing Company. First published in 1641.

Dogramaci, Sinan. 2020. "What Is the Function of Reasoning? Mercier and Sperber's Argumentative and Justificatory Theories," *Episteme*, 17, 316–30.

Evans, Jonathan St. B. T. 2010. *Thinking Twice: Two Minds in One Brain*, Oxford University Press.

Evans, Jonathan St. B. T., and Frankish, Keith, eds. 2009. *In Two Minds: Dual Processes and Beyond*, Oxford University Press.

Feldman, Richard. 2003. *Epistemology*, Prentice-Hall.

Foley, Richard. 1993. "What Am I to Believe?," in S. Wagner and R. Warner, eds., *Naturalism: A Critical Appraisal*, University of Notre Dame Press, 147–162.

Frances, Bryan. 2005. *Skepticism Comes Alive*, Oxford University Press.

Fumerton, Richard. 2006. *Epistemology*, Blackwell.

Gelman, Susan. 2003. *The Essential Child: Origins of Essentialism in Everyday Thought*, Oxford University Press.

Godfrey-Smith, Peter. 1996. *Complexity and the Function of Mind in Nature*, Cambridge University Press.

Goldman, Alvin. 1986. *Epistemology and Cognition*, Harvard University Press.

Goldman, Alvin. 1999a. *Knowledge in a Social World*, Oxford University Press.

Goldman, Alvin. 1999b. "Internalism Exposed," *Journal of Philosophy*, 96, 271–93.

Goldman, Alvin. 2005. "Kornblith's Naturalistic Epistemology," *Philosophy and Phenomenological Research*, 71, 403–10.

Goldman, Alvin. 2007. "Philosophical Intuitions: Their Target, Their Source, and Their Epistemic Status," *Grazer Philosophische Studien*, 74, 1–26.

Goldman, Alvin. 2012. *Reliabilism and Contemporary Epistemology: Essays*, Oxford University Press.

Goldman, Alvin, and McGrath, Matthew. 2015. *Epistemology: A Contemporary Introduction*, Oxford University Press.

Goldman, Alvin, and Pust, Joel. 1998. "Philosophical Theory and Intuitional Evidence," in Michael DePaul and William Ramsey, eds., *Rethinking Intuition: The Psychology of Intuition and Its Role in Philosophical Inquiry*, Rowman and Littlefield, 179–97.

Goldman, Alvin, and Whitcomb, Dennis, eds. 2011. *Social Epistemology: Essential Readings*, Oxford University Press.

Goodman, Nelson. 1973. *Fact, Fiction, and Forecast*, 3rd edition, Bobbs-Merrill.

Gopnik, Alison, Meltzoff, Andrew, and Kuhl, Patricia. 1999. *The Scientist in the Crib: Minds, Brains, and How Children Learn*, William Morrow.

Greco,John.2000.*PuttingSkepticsinTheirPlace:TheNatureofSkepticalArguments and Their Role in Philosophical Inquiry*, Cambridge University Press.

Harris, Paul. 2012. *Trusting What You're Told: How Children Learn from Others*, Harvard University Press.

Haugeland, John. 1985. *Artificial Intelligence: The Very Idea*, MIT Press.

Henrich, Joseph. 2016. *The Secret of Our Success: How Culture Is Driving Human Evolution, Domesticating Our Species, and Making Us Smarter*, Princeton University Press.

Howard, Ian. 2012. *Perceiving in Depth*, 3 vols., Oxford University Press.

Hume, David. 1888. *A Treatise of Human Nature*, edited by L. A. Selby-Bigge, Oxford University Press. First published in 1739.

Jackson, Frank. 1998. *From Metaphysics to Ethics: A Defence of Conceptual Analysis*, Oxford University Press.

Kahneman, Daniel. 2011. *Thinking, Fast and Slow*, Farrar, Strauss and Giroux.

Kahneman, Daniel, Slovic, Paul, and Tversky, Amos, eds. 1982. *Judgment under Uncertainty: Heuristics and Biases*, Cambridge University Press.

Kahneman, Daniel, and Tversky, Amos. 1972. "Subjective Probability: A Judgment of Representativeness," *Cognitive Psychology*, 3, 430–54; reprinted in Kahneman, Slovic, and Tversky. Page numbers in the text refer to the reprinted version, 32–47.

Keil, Frank. 1989. *Concepts, Kinds, and Conceptual Development*, MIT Press.

Kornblith, Hilary. 1993. *Inductive Inference and Its Natural Ground*, MIT Press.

Kornblith, Hilary, ed. 2001. *Epistemology: Internalism and Externalism*, Blackwell.

Kornblith, Hilary. 2002. *Knowledge and Its Place in Nature*, Oxford University Press.

Kornblith, Hilary. 2007. "Naturalism and Intuitions," *Grazer Philosophische Studien*, 74, 27–49.

Kornblith, Hilary. 2012. *On Reflection*, Oxford University Press.

Korsgaard, Christine. 1996. *The Sources of Normativity*, Cambridge University Press.

Kripke, Saul. 1980. *Naming and Necessity*, Harvard University Press.

Kunda, Ziva. 1999. *Social Cognition: Making Sense of People*, MIT Press.

Lackey, Jennifer. 2008. *Learning from Words: Testimony as a Source of Knowledge*, Oxford University Press.

Lackey, Jennifer, and Sosa, Ernest. 2006. *The Epistemology of Testimony*, Oxford University Press.

Lewis, Michael. 2017. *The Undoing Project: A Friendship That Changed Our Minds*, W. W. Norton.

Lord, Charles, Ross, Lee, and Lepper, Mark. 1979. "Biased Assimilation and Attitude Polarization: The Effects of Prior Theories on Subsequently Considered Evidence," *Journal of Personality and Social Psychology*, 37, 2098–109.

Maddy, Penelope. 2017. *What Do Philosophers Do? Skepticism and the Practice of Philosophy*, Oxford University Press.

Markman, Ellen. 1989. *Categorization and Naming in Children: Problems of Induction*, MIT Press.

Marr, David. 1982. *Vision*, W. H. Freeman.

Marušić, Berislav. 2015. *Evidence and Agency: Norms for Belief for Promising nad Resolving*, Oxford University Press.

Mercier, Hugo, and Sperber, Dan. 2011. "Why Do Humans Reason? Arguments for an Argumentative Theory," *Behavioral and Brain Sciences*, 34, 57–111.

Mercier, Hugo, and Sperber, Dan. 2017. *The Enigma of Reason*, Harvard University Press.

Millikan, Ruth. 1984. *Language, Thought, and Other Biological Categories: New Foundations for Realism*, MIT Press.

Millikan, Ruth. 1993. *White Queen Psychology and Other Essays for Alice*, MIT Press.

Moore, G. E. 1959. *Philosophical Papers*, Collier Books.

Moran, Richard. 2001. *Authority and Estrangement: An Essay on Self-Knowledge*, Princeton University Press.

Moser, Paul, ed. 1987. *A Priori Knowledge*, Oxford University Press.

Nagel, Jennifer. 2014. *Knowledge: A Very Short Introduction*, Oxford University Press.

Nisbett, Richard, and Borgida, Eugene. 1975. "Attribution and the Psychology of Prediction," *Journal of Personality and Social Psychology*, 32, 932–43.

Nisbett, Richard, Borgida, Eugene, Crandall, Rick, and Reed, Harvey. 1976. "Popular Induction: Information Is Not Necessarily Informative," in J. S. Carroll, and J. W. Payne, eds., *Cognition and Social Behavior*, Lawrence Erlbaum Associates; reprinted in Kahneman, Slovic, and Tversky. Page numbers in the text refer to the reprinted version, 101–116.

Nisbett, Richard, and Ross, Lee. 1980. *Human Inference: Strategies and Shortcomings of Social Judgment*, Prentice-Hall.

Nisbett, Richard, and Wilson, Timothy. 1977. "Telling More Than We Can Know: Verbal Reports on Mental Processes," *Psychological Review*, 84, 231–59.

Putnam, Hilary. 1975. *Mind, Language and Reality: Philosophical Papers*, vol. 2, Cambridge University Press.

Quine, W. V. O. 1969. *Ontological Relativity and Other Essays*, Columbia University Press.

Russell, Bertrand. 1912. *The Problems of Philosophy*, Oxford University Press.

Sellars, Wilfrid. 1963. *Science, Perception and Reality*, Routledge and Kegan Paul.

Sextus Empiricus. 1967. *Outlines of Pyrrhonism*, Loeb Classical Library, Harvard University Press.

Sosa, Ernest. 2007. *A Virtue Epistemology: Apt Belief and Reflective Knowledge*, vol. 1, Oxford University Press.

Sosa, Ernest. 2015. *Judgment and Agency*, Oxford University Press.

Sosa, Ernest. 2017. *Epistemology*, Princeton University Press.

Stein, Edward. 1996. *Without Good Reason: The Rationality Debate in Philosophy and Cognitive Science*, Oxford University Press.

Sterelny, Kim. 2003. *Thought in a Hostile World*, Blackwell.

Steup, Matthias. 1996. *An Introduction to Contemporary Epistemology*, Prentice-Hall.

Tomasello, Michael. 2014. *A Natural History of Human Thinking*, Harvard University Press.

Tomasello, Michael. 2019. *Becoming Human: A Theory of Ontogeny*, Harvard University Press.

Tversky, Amos, and Kahneman, Daniel. 1971. "Belief in the Law of Small Numbers," *Psychological Bulletin*, 2, 105–10; reprinted in Kahneman, Slovic, and Tversky. Page numbers in the text refer to the reprinted version, 23–31.

Wade, Nicholas, and Swanston, Mike. 2013. *Visual Perception: An Introduction*, Psychology Press.

Williams, Michael. 1996. *Unnatural Doubts: Epistemological Realism and the Basis of Skepticism*, Blackwell.

Williams, Michael. 2001. *Problems of Knowledge: A Critical Introduction to Epistemology*, Oxford University Press.

Williamson, Timothy. 2000. Knowledge and Its Limits, Oxford University Press.

Wittgenstein, Ludwig. 1969. *On Certainty*, translated by G. E. M. Anscombe and Denis Paul, edited by G. E. M. Anscombe and G. H. von Wright, Harper.

Wilson, Timothy. 2002. *Strangers to Ourselves: Discovering the Adaptive Unconscious*, Harvard University Press.

Zagzebski, Linda. 2009. *On Epistemology*, Wadsworth.

Index